PENGUIN BOOKS

SCIENCE AND PHILOSOPHY

Derek Gjertsen studied philosophy and the history of science at the universities of Leeds and Oxford. From 1960 to 1979 he taught at the University of Ghana. He is the author of *The Classics of Science: A Study of Twelve Enduring Scientific Works* and *The Newton Handbook*.

SCIENCE AND PHILOSOPHY

PAST AND PRESENT

DEREK GJERTSEN

Penguin Books

PENGUIN BOOKS

Published by the Penguin Group
Penguin Books Ltd, 27 Wrights Lane, London W8 5TZ, England
Penguin Books USA Inc., 375 Hudson Street, New York, New York 10014, USA
Penguin Books Australia Ltd, Ringwood, Victoria, Australia
Penguin Books Canada Ltd, 10 Alcorn Avenue, Toronto, Ontario, Canada M4V 3B2
Penguin Books (NZ) Ltd, 182–190 Wairau Road, Auckland 10, New Zealand

Penguin Books Ltd, Registered Offices: Harmondsworth, Middlesex, England

First published in Pelican Books 1989
Reprinted in Penguin Books 1992
1 3 5 7 9 10 8 6 4 2

Printed in England by Clays Ltd, St Ives plc
Set in Linotron Ehrhardt

For Allan and Peggy

CONTENTS

INTRODUCTION

POLYWATER:

1st-2nd-3rd editions	1942–64	No entry
4th edition	1971	Anomalous water, name suggested for a form of water which can be prepared by condensing ordinary water in fine quartz or glass capillary tubes. Polywater has a density of about 1.4 gm per cc and remains stable up to about 500° C. It also has a higher viscosity than ordinary water.
5th edition	1979	Anomalous water. A reported form of water differing in properties . . . from normal water.
6th edition	1986	Anomalous water . . . It is now accepted that these properties were due to . . . impurities rather than to any differences in the molecular structure of the water itself.

The Penguin Dictionary of Science (1942–86)

The theme of the present work is the manner in which science and philosophy have interacted since the time in the sixth century BC when the sage Thales declared everything to be water, and thereby, most implausibly, founded both disciplines. Today the disciplines of science and philosophy seem to be so far apart, so fundamentally distinct, that it is initially hard to see how they could be related in any significant manner. Scientists synthesise steroids, report on the geology of distant planets, manipulate individual neurons, explore the detailed structure of atomic nuclei, and spend billions of pounds within their palatial laboratories. In contrast philosophers, tucked away in cramped offices, are seen as endlessly exploring texts of Plato and Aristotle, or pursuing such trivial and purely linguistic issues as the exact nature of the difference between kindness and kindliness.[1] And while most philosophers may treat science with immense respect, many scientists have only, according to Peter Medawar, an 'exasperated contempt' for philosophy.[2]

Things were not always so, and it would be well to begin by demonstrating some aspects of an earlier intimacy. The most obvious signs of a closer relationship are to be found in language. Many have been struck by the late appearance in English of the word *scientist.* In 1840 William Whewell noted that there was no simple and natural way to refer to 'a cultivator of science in general'. He was, he concluded, inclined to call him 'a scientist'.[3] Before Whewell scientists tended to refer to each other as philosophers, or, more fully, as natural philosophers. For this reason Newton's lengthy treatise on mathematical physics was given the title *The Mathematical Principles of Natural Philosophy* (1687). More informally, as in a 1679 letter to Robert Hooke, Newton declined to discuss a number of scientific issues on the ground that he had 'no time to entertain philosophical meditations'.[4]

The usage persisted. In 1807, for example, the physicist Thomas Young chose to publish his Royal Institution lectures under the title *Lectures in Natural Philosophy.* A year later Dalton introduced the modern concept of the atom into science in a work entitled *A New System of Chemical Philosophy.* Further signs of confusion can still be seen in the titles of a number of academic chairs. Professors of physics at Oxford and Cambridge hold such anachronistically named chairs as the Dr Lees Professorship of Experimental Philosophy, and the Jacksonian Professorship of Natural and Experimental Philosophy. Perversely, Cambridge professors of philosophy occupy chairs of moral and mental science.

Yet more evidence comes from the lives of leading scientists and philosophers. Most dramatic of all is the presence of a number of scholars, admittedly few, who spanned both disciplines and rank today as both leading scientists and leading philosophers. A strong case can be made, for example, for presenting Descartes as the founder of both modern science and modern philosophy. Equally eminent in both ancient philosophy and science was the dominating figure of Aristotle. Other scholars as likely to be found in histories of science as in histories of philosophy include Pascal, Leibniz, Boyle, Newton, Whewell, Mach, and Bertrand Russell.

Many other important philosophers, while never dominating the science of their day, have nevertheless worked for part of their lives as professional scientists. Locke and Lotze were both physicians; Whitehead, Frege, and Ramsey were mathematicians; Wittgenstein was an engineer; Michael Polanyi a physical chemist, Schlick a physicist, James Ward a physiologist, and William James a psychologist. In more recent times a large number of philosophers, though never actively pursuing a career in science, have all – like C. D. Broad, Karl Popper, S. E. Toulmin, W. V. Quine, Rudolf Carnap – taken first degrees in maths or physics.

There are also several leading scientists, some Nobel laureates, who while following their scientific careers concern themselves with philosophical issues. Werner Heisenberg, for example, one of the founders of modern quantum theory, has written extensively on the philosophical foundations of modern physics. David Bohm, another theoretical physicist, has argued in a number of works for a philosophical framework within which some of the more startling aspects of quantum theory would appear less paradoxical. In addition there are the various books of the immunologist Peter Medawar, in which, much influenced by the philosopher Karl Popper, he has attempted to display the logical structure of science.[5]

Philosophers in their turn have been much influenced by science, and, indeed, for some the source of their initial philosophical awakening, and their continued inspiration, has been solidly located within science itself. One who has left a full account of his philosophical basis, along with much else, is the Victorian philosopher Herbert Spencer. Born in 1820, Spencer began his career as an engineer, but, with the aid of a small inheritance, he turned in later life to writing and scholarship. He recorded in his *Autobiography* (1904) how he had come across the theory of Karl von Baer that every plant and animal in its development changes 'from homogeneity to heterogeneity'.[6] Spencer found the thought so impressive that he spent the rest of his long life showing the relevance of the principle to any study of society, science, human development and thought, and to much more besides. It took him some thirty years and ten large volumes of his *Synthetic Philosophy* to exhaust all aspects of von Baer's insight.[7] Asked about the impact of von Baer on his work, Spencer replied in characteristically prolix terms: 'If anyone says that had von Baer never written I should not be doing that which I now am, I have nothing to say to the contrary – I should reply it is highly probable.'[8]

Popper is another philosopher who has attributed his intellectual awakening to science. Unlike Spencer, however, it was not a particular theory which so impressed Popper, but rather an approach he saw in science, so distinctive as to separate it from all other disciplines. In the Vienna of the post-war years Popper was initially attracted to the theories of Marx, Freud, and Adler. Nothing seemed to lie outside their scope, and explanations were instantly provided to deal with any new problem.[9] Against this, he noted, stood the precise prediction derived by Einstein from his General Theory, so triumphantly confirmed by Eddington's observations of the 1919 solar eclipse:

This, I felt was the true scientific attitude. It was utterly different from the dogmatic attitude which constantly claimed to find 'verifications' for its favourite theories. Thus I arrived at the end of 1919 at the conclusion that the scientific attitude was the critical attitude, which did not look for verifications but for crucial tests; tests which could *refute* the theory tested, though they could never establish it.[10]

Like Spencer before him, Popper has spent the rest of his life defending, and pursuing the implications of, his central insight.

It must of course be acknowledged that there are many philosophers who owe little to science. G. E. Moore, for example, the leading figure of Cambridge analytical philosophy, in talking of his own intellectual roots, noted that 'I do not think that the world or the sciences would ever have suggested to me any philosophical problems. What has suggested philosophical problems to me is the things other philosophers have said about the world or the sciences.'[11] Thus,

Moore reported, he had been interested mainly in two kinds of problem. First, the problem of trying 'to get really clear as to what on earth a given philosopher meant by something which he said', and secondly, determining whether there were any 'satisfactory reasons . . . for supposing what he meant was true, or, alternatively, was false'.

Philosophers, however, from Socrates onwards, have seen science as something more than a convenient source of problems and insights. For many its very existence has been a challenge, calling for definition, analysis, and justification. Three questions in particular have long puzzled philosophers. There is first the demarcation problem and its concern with the features which supposedly distinguish science from philosophy and all other intellectual disciplines. Secondly there are a number of methodological issues centred ultimately on attempts to identify and describe the basic features of the scientific method. And, finally, there are questions of justification as they arise from consideration of the traditional problem of induction.

Yet, in recent times, a new dimension has been added to the subject, and the philosophical study of science has consequently reached an exciting time in its development. When I was first introduced to the subject in the 1950s the influence of the logical positivists continued to dominate the field. Problems were invariably reductive and logical. Much effort was expended on considering the status of the laws of nature and of theoretical entities, and of how they could be both rigorously defined within a deductive system and still remain empirically meaningful. Some remarkable conclusions were reached. Electrons ceased to exist and became logical constructions created out of meter readings and tracks on photographic plates.[12] Laws of nature were rejected as empirical truths and described instead as rules for finding our way about reality.[13] The problem of induction was transformed into a new discipline of confirmation theory and, amidst a plethora of paradoxes, much time was spent by inductive logicians considering the logical complexities of the proposition 'All crows are black.'[14]

A few philosophers, however, stood to one side, claiming that they could recognise little of science in these formal manipulations. They pursued instead in a somewhat lonely fashion their own idiosyncratic interests. Historians of science like Alexandre Koyré, for example, and philosophers like Stephen Toulmin, argued for and exhibited a less formal and more historical approach to the subject. From such works as Koyré's *From the Closed World to the Infinite Universe* (1957) philosophers could begin to learn about the multiple sources of the scientific revolution – that it contained elements derived from theology and philosophy, as well as from experiment and observation. Also displayed, although initially as no more than a sub-theme, were the facts of scientific change. Few, however, were prepared to abandon the positivists' obsession with the supposed 'structure' of science for more substantive matters. Caught in a no-man's-land, the few early refugees from positivism found themselves dismissed by historians as superficial, and ignored by philosophers as irrelevant.

And so matters rested until 1962 and the appearance of T. S. Kuhn's remarkable work, *The Structure of Scientific Revolutions*. At a stroke the issue of scientific change had been placed at the centre of debate for historians and philosophers of science alike. Historians began to document the facts of scientific change as they were revealed in numerous past revolutions, and philosophers began to assess the implications of this work.

They were soon joined by a number of radical sociologists who quickly saw within the Kuhnian legacy the seeds of a new and more cogent relativism, from which not even science could remain exempt.[15] The realisation began to haunt post-Kuhnian philosophy. The inevitability of scientific revolutions presumably meant that, just as the theories of the past had been discarded, current accounts of nature would eventually face a similar fate. And, given the rate of change of modern science, this fate was unlikely to be long delayed. The implications of this view were startling. The papers contained in any recent issue of *Nature* were apparently no more reliable than accounts of nature published long ago by Aristotle and Descartes. Nobel prizes, it seemed, were being awarded for bogus contributions to knowledge, and billions of dollars were being spent needlessly on the detection of fictitious particles.

Problems of historical change had thus led a new generation of philosophers to the impasse which had earlier trapped the logical positivists. Whereas the positivists had failed to show how experience could ever justify many important propositions of science, the post-Kuhnian epistemologist was beginning to find it equally difficult to accept as true the propositions of science, however well supported

th...									d the widely
ac...									...re of young
T...									replace them
w...

a...									...ies, and un-
t...									...tist continues
n...									...nd build ever
t...									d successes of

s...									...ny account of
...									truth will be
...									ter 11. Before
...									ers will first be
...									work out some
...									When and how
...									Having emerged
...									the distinction?
...									will be shown in
...									distinction prove
unworkable. In the following two chapters questions about the
origins and conclusions of philosophical and scientific problems will
be examined. It is often claimed that many scientific problems have
their roots within philosophy; it is less often demonstrated what these
roots are and how they have led to science. Another claim frequently
made is that philosophical attempts to solve important problems are
essentially exploratory and seldom successful. The ultimate fate for
all philosophical problems is, therefore, either to be pursued endlessly by philosophers, or to be absorbed painlessly into science where
eventually they find an appropriate solution. Both claims are shown
to be largely without foundation.

In Part III the issue of method is raised. Scientific methodology
has long presented philosophers with both a challenge and a temptation. The challenge, the subject of chapter 6, is to describe as
accurately as possible the actual methodology of the sciences. The
temptation, covered in the following chapter, is to adopt precisely
these methods, such as they are, for the solution of philosophical
problems. Many have been tempted by the thought that if only
they could work out how to adapt the methodology of science to

philosophical ends, then philosophy could become as successful as science. The fate of such attempts is described in chapter 7.

But philosophers have seldom been content merely to describe and to learn from science; they have also sought to challenge it directly and offer philosophical refutations of scientific positions. While frontal assaults of this kind can be embarrassingly inept, they can also be rich in insight. Philosophical challenges to science can come in two styles; they form the subjects of chapters 8 and 9.

In the final Part, V, the more difficult problems of the validity of science and its epistemological base are considered. In chapter 10 it is argued that science, in the form we understand it today, had to face in earlier times alternative ways of understanding nature. The fact that we now seek to understand nature in scientific terms is the result of battles fought long ago, and of battles won at least partly through philosophical effort. Something of this distant battle will be described. And finally, in chapter 11 the question I began with, the difficulty of reconciling truth and change within science, will be answered.

Before I begin, however, I must apologise over a couple of stylistic matters. Throughout the book I refer to philosophers as if they are exclusively male. The reason my text is scattered with 'he's and 'him's is simply that I know no way to avoid this without resorting to an unacceptable prolixity or without adopting my own incomprehensible neologisms. I also refer throughout the text to the subject of philosophy, although I have in mind only the more restricted field of the philosophy of science as it is practised within the analytical tradition of western thought. I am well aware that there are other traditions, some of importance. None, however, as far as I can see, challenge the views expressed below on the nature of science and philosophy.

Pascal warned against authors talking of 'my book' in the manner of 'bourgeois homeowners, with "my house" always on their lips'. Rather, he advised, 'They should speak of our book, our commentary, our history, etc., since, generally speaking, there is far more in them of others than of their own.' The sentiment is well known to writers of synoptic works on science and philosophy. I have learned a good deal from many writers, and where I have borrowed directly I have sought to make the appropriate acknowledgement. Often, however, the borrowing, though no less real, has been more indirect. And while they may not recognise, nor agree with, my use of their work, a further acknowledgement is called for. Much of my approach to the philosophy of science is based on the work of my first teacher in the subject, Stephen Toulmin. I have also learned much from the writings of Ian Hacking, Larry Laudan, Robin Horton, the still deeply puzzling work of Thomas Kuhn, and the ever-rewarding publications of the historian of science, Alexandre Koyré. Finally, I would like to thank once more Ted Honderich for his general support and for his helpful comments on an earlier version of the work. The errors, infelicities, irrelevancies, and obscurities remaining are, I fear, entirely my own responsibility.

PART I

EMERGENCE AND DIVERGENCE

WHEN DID PHILOSOPHY AND SCIENCE DIVERGE?

> But Socrates first called philosophy down from the skies [philo-
> sophia de caelo devocata], set it in the cities and even introduced it
> into homes, and compelled it to consider life and morals, good and
> evil.
>
> Cicero, *Tusculan Disputations*

It is widely held, as by Ayer, that it was not until 'the nineteenth
century that the paths of philosophy and science began to diverge'.[1]
The change was supposedly linked with 'the multiplication of scientific knowledge, and the consequent increase of specialisation'. More
recently Richard Rorty has claimed that it was not until after Kant
'that our modern philosophy–science distinction took hold'. Before
then philosophers fought to distinguish themselves from theologians.
Only when this victory had been gained, Rorty insists, could they
begin to ask how their work differed from science. Realising that
metaphysics – the description of the heavens and the earth – had
been pre-empted by science, Kant and his successors would substitute the study of epistemology in the sense of 'a theory distinct from
the sciences because it was their *foundation*'.[2]

There is little doubt that the Ayer–Rorty dating is seriously wrong.
By the late seventeenth century philosophers such as Locke had
already begun to recognise their own role as secondary and deriva-
tive. In a famous passage prefacing his *Essay Concerning Human
Understanding* (1690), he noted that not everyone could hope to be a
Boyle, Sydenham, Huyghens, or Newton; for such as himself it was

ambition enough 'to be employed as an under-labourer in clearing the ground a little, and removing some of the rubbish that lies in the way to knowledge'.

Locke had shortly before made a considerable effort to understand Newton's work. He approached the Dutch physicist Huyghens on how best, given his lack of mathematics, to tackle *Principia*. Newton went so far as to prepare for him a special *Demonstration* of one of his central theses that 'the Planets by their gravity towards the Sun, may move in Ellipses'.[3] As a result of his efforts Locke became known to his contemporaries as, in Desaguliers's revealing phrase, 'the first Newtonian philosopher without the help of geometry'.

The distinction can be pushed back much further, as far in fact as antiquity. The term philosophy itself was, according to Diogenes Laertius, first used by Pythagoras in the sixth century BC: 'Life, he said, is like a festival; just as some come . . . to compete, some to ply their trade but the rest come as spectators, so in life, the slavish men go hunting for fame or gain, the philosophers for the truth.'[4] What kind of philosophy did Pythagoras pursue? According to Cicero, in his *Tusculan Disputations*, he spoke about 'number and movement, and the source from which all things arise and to which they return; and these early thinkers inquired zealously into magnitude, intervals, and courses of the stars, and all celestial matters.'[5] To a modern reader Pythagoras and his contemporaries seem to have been concerned exclusively with problems of astronomy, physics, and mathematics. G. E. R. Lloyd has noted how repeatedly the Presocratics discussed such questions as 'the causes of lightning, thunder, earthquakes, comets and stars',[6] i.e. natural phenomena.

Their concerns, we can now recognise, were those of science. They had not, however, by then progressed far in developing the essential scientific procedures of experiment, observation, and calculation. In their place they tended to resort, in Lloyd's phrase, to 'drawing analogies with familiar objects'. Typical examples would be the claim of Empedocles that the sea is 'the sweat of the earth', or Anaximenes's comparison of lightning with the flashes made by oars as they cleave the water, or finally, the chariot wheel Anaximander used as a model of the heavens.

Most ancient sources agree that philosophy, as we now understand it, emerged as a reaction to the poverty of this early scientific approach. This remarkable intellectual revolution occurred in the fifth century BC and was initiated by Socrates. Aristotle, for example, noted that at this time 'The study of nature was given up, and philosophers turned their attention to practical goodness and political science.'[7] Cicero, too, recorded that it was Socrates who, in a memorable phrase, 'first called philosophy down from the sky' and turned it to the consideration of 'life and morals, good and evil'.[8]

A fairly full account of this revolution can be found in Xenophon's *Memoirs of Socrates*.[9] The picture presented is an unfamiliar one in that it shows Socrates castigating the science of his day as confused and valueless. Four arguments were offered and, in one form or another, they have continued to be raised by critics of science down through the centuries.

1   How could anyone, Socrates asked, hope to understand nature without first gaining knowledge of himself? The point was also made by Plato, who attributed to Socrates the following argument: 'I cannot yet, in the words of the Delphic precept, "know myself" and it seems to me ridiculous to be studying alien matter when still ignorant of this.'[10] Two thousand years later, in the Introduction to his *Treatise* (1738), Hume also would argue that science was dependent on human nature, and that scientific progress would only become possible when we were 'thoroughly acquainted with the extent and force of the human understanding'. Both Socrates and Hume were clearly denying the autonomy of science. They were further insisting that science, though not derivative, was at least dependent upon philosophy.

2   Confirmation for his view, Socrates claimed, was provided by the needless disputes characteristic of the science of his day. Whenever confident claims were made by one scholar they would be opposed by equally confident but conflicting views from hostile critics. Socrates noted that while some argued that 'reality is one', others insisted that it is 'infinitely many'; and while some claimed that 'everything is always in motion', others held that 'nothing can ever be moved.' Once again the views were echoed by Hume. He too was struck by the fact that in the science of his time: 'There is nothing which is not the subject of debate, and in which men of learning are not of contrary opinions.'

3 Surprisingly, for a supposedly theoretical Greek philosopher,
   Socrates went on to object that science had little practical value.
   Will a knowledge of nature enable us 'to produce at will winds
   and rain'? Astronomy was of some value, but only in so far as it
   enabled us to tell the time at night, and the seasons of the year.
   Even the Greek passion for geometry was denigrated. Learn as
   much, he advised, as will enable you 'to apportion land accurately
   in point of measurement'. The remainder was best ignored, as it
   was not only useless, but 'capable of wasting a man's life and
   debarring him from learning many other useful things'.

4 His final objection to science, however, finds few modern echoes.
   The gods, he claimed, would be unlikely to be pleased by
   attempts to pry into matters they had chosen to conceal. Far
   better not to enquire too closely into the manner in which the
   gods regulate the heavens lest, like Anaxagoras, we be driven
   mad. We would be better employed studying divination.

By the time of Socrates a clear distinction had been drawn between
the study of nature and the study of man. The distinction was picked
up and developed by Aristotle. All intellectual activity, he argued in
the *Metaphysics*, was either productive, practical, or speculative.[11]
The productive disciplines, involving a variety of arts and crafts, lie
outside our scope. The practical disciplines – ethics, politics, and
economics – were all clearly concerned with man. Inevitably, the
speculative disciplines – maths, physics, and metaphysics – dealt with
nature.

There is more to man, however, than his relationship with the
state, the economy, and his family. What of his rational powers?[12]
According, once more, to Diogenes Laertius, it was Zeno of Citium
(335–263 BC), the founder of the Stoics, who first enlarged the
traditional division of learning to include logic: 'They liken philosophy to an animal, comparing logic to the bones and sinews, ethics
to the flesh, and physics to the soul.'[13]

Something very like this broad Stoic division of knowledge survived the medieval period unscathed.[14] Thus, Hugh of St Victor in
the twelfth century distinguished between theoretical, practical,
logical, and mechanical philosophy. The basic division between
human and natural philosophy was preserved by Bacon in his

Advancement of Learning (1605). While natural philosophy contained such subjects as physics, geometry, engineering, and astronomy, Bacon assigned the contrasting disciplines of logic, rhetoric, morals, and government to human philosophy. Towards the end of the century Locke concluded his *Essay* (1690) with a reiteration of the Stoic tripartite division of knowledge into physics, 'the knowledge of things as they are', practica, 'the skill of right applying our own powers and actions for the attainment of things good and useful', and, thirdly, semiotic or logic.

Even the elaborate division of knowledge contributed to the *Encyclopédie* (1751–76) by Diderot and D'Alembert followed the traditional system in outline.[15] The scheme began by dividing the understanding into the three faculties of *memory*, concerned with all aspects of history, *imagination*, devoted to poetry, and *reason*. This last faculty was identified with philosophy, which in turn, and ignoring all subdivisions, was broken down into the following broad categories:

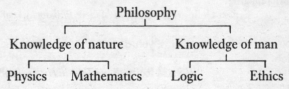

The detailed terminology and divisions appear strange. Botany, for example, is seen as part of physics, and military architecture finds itself under ethics. Despite this it is possible to recognise the broad divisions of learning still in force today. At one side such subjects as optics, pneumatics, and mechanics form part of the study of nature, while other subjects like perception, judgement, and reasoning form part of the study of man. Something of a return to this ancient tradition can be seen in modern universities. At one time alongside the Faculty of Science there could invariably be found the curiously named Faculty of Arts. While the science faculties have seen no reason to rename themselves, many arts faculties have decided to adopt the more appropriate title of Humanities Faculty.

Having identified the moment of divergence, it remains to characterise the nature of this intellectual schism. How, in fact, do science and philosophy differ?

2

HOW DO PHILOSOPHY AND SCIENCE DIFFER?

If all American philosophers were gathered into one place and an atomic bomb exploded [on them], American society would remain totally unaffected. No one would notice any difference, and there would be no gap, no vacuum in the intellectual economy that would require plugging.

Ernest Gellner, *The Devil in Modern Philosophy* (1984)

If you can synthesise steroids, you do not need historical legitimation.

Richard Rorty, *Philosophy in History* (1984)

Having shown that science and philosophy have long diverged, we must now identify the nature of this divergence. Many attempts have been made to distinguish between philosophy and science by appealing to a few simple criteria. They invariably end in failure and confusion. Both science and philosophy are complex and far from uniform disciplines, with long histories and varying aspirations. They are unlikely to be distinguished in a phrase or two. Yet, from Aristotle to Popper, philosophers have persisted in their unrealistic attempts.

Consider, for example, Roger Scruton's recent account of science as 'the realm of empirical investigations; it stems from the attempt to understand the world as we perceive it, to predict and explain observable events and to formulate the laws of nature (if there be any).'[1] Let us take Scruton's proposed criteria one by one.

Is science really 'the realm of empirical investigations'? What, if so, of the enormous areas of empirical investigation lying completely outside science? Someone checking the authenticity of a supposed first edition of *Oliver Twist*, the identity of his wife's lover, the soundness of a company's shares, or the reliability of a new watch, will employ a variety of investigative techniques, mostly empirical, none of which singly or collectively constitute science. Scruton's proposal succeeds in transforming the cricketer's guide *Wisden* and the telephone directory into basic scientific texts.

There are also many scientists who seldom, if ever, pursue empirical investigations. Mathematicians and mathematical physicists, for example, frequently ignore such concerns. In 1543 Copernicus allowed his *De Revolutionibus* finally to appear. It contained his arguments for the replacement of the earth-centred solar system of antiquity with his claim that the sun was the centre of the universe and around it all other celestial bodies moved in circular orbits. One will look in vain through this large work for the empirical investigation on which his conclusions were based. Instead, a number of philosophical and mathematical arguments are presented to show the superiority of his views. The most obvious point, and one well known to Copernicus, that predictions based on the Ptolemaic system were often wrong by a week or more, he simply ignored. Largely indifferent to empirical questions of this kind, Copernicus was much happier when considering not the accuracy of the system but its purity and harmony.

Further, does science really aim at establishing laws of nature? One difficulty is that there are well-established parts of science in which laws of nature play virtually no role at all. In evolutionary biology, for example, laws are seldom met. The few that have been suggested, such as Dollo's law on the irreversibility of evolution, seem to be little more than useful generalisations. While other laws, such as Cope's claim that organisms become larger as lineages evolve, have been shown to be 'by no means universal or irreversible'.[2]

Equally damaging is the point that scientists coped for centuries without benefit of law before Descartes introduced the notion into science. It is surprising to discover that Copernicus, Galileo, and Gilbert, along with other founders of the scientific revolution, make no reference to scientific laws. Galileo spoke of principles, ratios, and proportions, while Copernicus adopted an earlier terminology and spoke approvingly of the harmonies and symmetries he saw in nature. Even the familiar three laws of planetary motion discovered by

Kepler between 1609 and 1619 were not presented by him unambiguously as laws. It was left to Descartes, Boyle, and Newton later in the century to develop a clearer notion of the concept.[3]

Finally, what of Scruton's reference to prediction and explanation in science? Many outside science are, of course, dominated by these considerations. The gambler, stock market tipster, and the astrologer, are as much concerned with these matters as any scientist. There are also areas of science, palaeontology and systematics, for example, where the ability to describe and classify are more valued than the power to explain and predict.

Nor would much be gained by replacing Scruton's account with an alternative set of features. Talk of experiment and observation, or verification and falsification, are likely to suffer the same fate as Scruton's 'prediction and explanation'. The reason for this is quite clear, and has already been mentioned: complex, wide-ranging, and far from uniform systems of thought with long and varied histories are unlikely to be captured within the confines of a slogan. Yet, if the slogans do not establish dichotomies, they do at least point to areas of divergence. The aim in what follows will be, therefore, to identify some of these divergent areas, and to seek to establish how sharp, radical, or superficial they really are. In most cases it will be found that though it is indeed possible to distinguish between science and philosophy, the distinction can often be far from sharp, and is frequently one of degree rather than substance.

Armchairs and laboratory benches

Philosophy is frequently diagnosed as being no more than an armchair activity. Compared to scientists vigorously investigating nature in the field or laboratory, the philosopher, from the confines of his well-padded armchair, in his comfortable ivory tower, must find himself restricted to considering the writings of other armchair-bound philosophers.

It should of course be realised that scientists themselves are no strangers to armchairs. J. J. Thomson, for example, discoverer of the electron, was reported to have spent 'a good part of most days in the armchair of Maxwell doing mathematics'.[4] Another physicist, Paul Dirac, noticed that much of his life has been spent 'just playing with equations and seeing what they give'.[5]

Perhaps, however, the lifestyles of mathematical physicists are unrepresentative. More typical would be the career of a naturalist like Darwin. A four-year circumnavigation, during which he examined the geology, flora and fauna of wherever he visited, was followed by a lifetime of continuous observation and experiment. Despite poor health, a large family, and the writing of numerous books, Darwin worked on such varied phenomena as vegetable mould, the fertilisation of orchids, earth worms, insectivorous and climbing plants, and, besides much more, an exhausting survey of barnacles.

Yet, it is tempting to ask, could not the theory of evolution have emerged from the armchair of a well-read scholar? Oddly enough, it could, and it emerged in this manner not once but twice. As is well known, the theory of natural selection was developed quite independently by both Charles Darwin and Alfred Wallace. In 1858 Wallace was pursuing his career as a bug-hunter on the island of Ternate in the Moluccas. Suffering from malaria, and with not an armchair in sight, Wallace was forced 'to lie down for several hours'; with nothing to do but 'to think over any subjects then particularly interesting me', he began to muse upon the thesis of Malthus.[6] Eventually, it 'flashed' upon Wallace that 'the fittest would survive', and with this central insight he began to construct the modern theory of evolution.[7] Some twenty years earlier, Darwin, from the armchair of his London home, had arrived at a similar conclusion. He had happened to read for amusement 'Malthus on population'. It suddenly 'struck' him that it contained a process leading to 'the formation of new species'.[8]

The examples of Darwin and Wallace clearly reveal the obvious but frequently ignored point that much of science involves no more than thinking. And thinking, of course, can take place anywhere – the bathroom of Archimedes, the garden of Newton, a bus travelling to Clapham for the organic chemist Kekulé, or the armchairs of Darwin and Maxwell.[9] Inevitably, there is much more to science than merely thinking about the world. Ideas once conceived have to be worked out, tested, analysed, and examined in numerous ways. The process, no doubt, is a complex one and will involve at some point a certain amount of observation and experimentation. The point remains however that, in part, science is as sedentary and cerebral an activity as academic philosophy.

Scholars can accordingly study science as a system of *thought*. It is, of course, a system of informed thought; it is, also, a system of thought constrained by the scientific assumptions and theories of the day. One of the best-known practitioners of this approach has been Alexandre Koyré. In, for example, his aptly named *From the Closed World to the Infinite Universe* (1957), Koyré demonstrated how, between Nicholas of Cusa in the fifteenth century and Newton in the late seventeenth, science moved, almost as an exercise in philosophy, from the former to the latter commitment.

What, though, of the more empirically based and descriptive sciences? Much can, in fact, be gathered from a suitably sited armchair. Some biologists do indeed spend the bulk of their career buried in the bowels of a museum, and never, professionally, see the light of day. Linnaeus and Buffon in the eighteenth century constructed the natural history of the known world from their studies in Uppsala and Paris. To them came samples, specimens, and descriptions from travellers, collectors, students, and other scholars.

Nor are such sedentary pursuits the exclusive preserve of the past. Even more literary has been the work of the Chicago palaeontologist, John Sepkoski. Geologists have long been aware that the fossil record reveals evidence of periodic mass extinctions. Best known is the disappearance of the dinosaurs at the end of the Cretaceous. More extensive, though less well known, was the mass extinction at the end of the Permian when as many as 96 per cent of marine species were eliminated. Numerous other catastrophic periods have been identified by geologists.

The temptation to speculate on the causes of such mass extinctions is, of course, irresistible. And geologists have, consequently, frequently conjured up whole series of passing comets, explosive volcanoes, or massive meteorites. Before engaging in the search for causes, Sepkoski argued, it would be as well to collect much more data about the process of extinction. He began to compile *A Compendium of Fossil Marine Families* (1982) in which the age of first and last occurrences of some 3,500 taxa are listed. The work has been regularly revised on computer and is to be followed by a list of genera. Sepkoski's field area, according to one colleague, 'is the library and its thousands of reports and monographs that record fossils found since the middle of the eighteenth century'.[10] Although almost entirely literary and sedentary, Sepkoski's work has none the less been highly suggestive. Statistical analysis of the *Compendium* data by Sepkoski and his colleague David Raup suggested to them that over the past 250 million years extinctions had been neither continuous nor random. Rather, they had struck with a periodicity of 26 million years. Palaeontologists are still evaluating the claim.

Yet, however much science of this kind is carried out in the stacks of libraries and museums, many will still seek to distinguish it in a very simple way from the work of armchair philosophers. Scientists like Sepkoski may indeed, in part of their work, spend much time in their chairs. But this is only because from this point they can more easily consult the publications of scientists who have dug the soil, fished the seas, and explored extensively the strata of the world. Unlike philosophers they are relying, not on their own speculative imagination, but on the empirical labours of their fellow geologists.

But, equally, it can be argued, political and moral philosophers, as well as philosophers of law, history or science, seldom rely exclusively on the power of their imaginations. An examination of Hume's writings on morals, politics, and religion, for example, will reveal that, within the confines of his time, they depend on material gathered from classical authors, from history and literature, and from the growing corpus of travellers' descriptions of distant lands and cultures. Thus, he began his *The Natural History of Religion* with the claim that polytheism arose not from 'a contemplation of the works of nature', but from a concern with 'the events of life'.[11] Evidence was presented from Pliny, Hesiod, and Aristotle of the manner in which the ancients assigned specific duties to their deities. The Greeks even had, Hume noted, 'a God of sneezing'.[12] When he turned to argue that deities need not be taken as 'Creators of or Formers of the World', evidence was produced from Regnard's *Voiage de Laponie* that the Laplanders worship any 'large stone which they meet with an extraordinary shape'.[13]

It is difficult to see how Hume's procedures at this point differ from the accounts Linnaeus and Buffon, his contemporaries, offered of the primates. They too quoted Pliny, Aristotle, and other classical sources; they too relied upon the more recent accounts of travellers. And, of course, they too laboured from the security of their armchairs. Hume's views may even have been the more accurate.[14]

Generality

Asked to identify a characteristic feature of philosophy, many philosophers appeal to the notion of generality. In 1914 Russell asked what features distinguish 'the province of philosophy from that of the special sciences', and went on to note that in the first place 'a philosophical problem must be general', and philosophical propositions 'must be applicable to everything that exists or may exist'.[15] Russell has been followed by A. J. Ayer with his claim that philosophy differs from science 'in the greater generality of the questions it raises'.[16] For example, while physicists, psychologists, and historians may all reject any number of arguments as invalid, the concern of the philosopher is not to assess whether or not a particular piece of evidence is adequate but to demand 'a justification for any form of non-demonstrative argument'.

The position can be traced to Aristotle. In his *Metaphysics* he spoke of a science which 'investigates being *qua* being and its essential attributes'. It differed from all the 'special sciences' in that it did not single out one aspect of being and study the essential attributes of that one part.[17] That science was, of course, metaphysics.

In this manner it is natural to suggest that philosophers are concerned with such very general concepts as cause, explanation, validity, rights, duties, thought, and existence. Lawyers, for example, could be concerned with legal rights, politicians with political rights, authors with copyrights, and theologians with natural rights; philosophers, however, according to this view, should be concerned with rights as such, rights without restriction or qualification.

Some philosophical works do display this feature. Quine's claim that 'to be is to be the value of a bound variable', or Berkeley's comparable claim that 'to be is to be perceived', are clearly very general statements about existence or being. Equally, Gilbert Ryle's *The Concept of Mind* (London, 1949) was genuinely about the notion of mind, any mind, while Sir Peter Strawson's *Individuals* (London, 1959) sought to be just as unrestricted about the notion of an individual substance.

Yet, many other philosophical works seem to aim for and to be satisfied with much more restricted domains. While all philosophers would be delighted to be able to give an account of existence in general, most recognise that it would be triumph enough if they could give a plausible account of mathematical existence alone, or theoretical existence, or even bodily existence. Again, there are excellent studies of legal causation which completely ignore problems concerned with the role of causal laws in quantum physics, in biology, and in theology. Indeed, philosophical accounts of causation which apply to quanta, the law, and miracles are far from plentiful.

Other philosophical works have dealt with a number of detailed and quite specific issues. At one time philosophical logicians were much concerned with the status of the proposition 'Nothing can be red and green all over'; at other times such specific questions as whether computers can think, or whether a private language could ever be invented, have been much discussed. It could of course be argued that the implications are far-reaching, and often concluding paragraphs are added pointing out the implications of the argument for the more general issues. In a similar manner scientific articles often conclude their accounts of their detailed and incredibly expensive work with a passing reference to the implications their results could have for clearer understanding of heart disease, cancer, or some other acceptable outcome. The sincerity of both conclusions should not be taken too seriously.

It must also be accepted that some scientific research is concerned with questions as general as any are ever likely to be. The physicist's account of motion, space, time, order, and waves seems to lack little in generality. As does the biologist's discussion of life and development. Compared with these the philosopher's account of political sovereignty and moral obligation seems positively parochial. There seems little to be gained by attempting to establish a distinction which belongs more to the rhetoric than to the substance of the situation.

Views of the past

Philosophy and science also differ in the views they take of their own pasts. As Jonathan Ree has noted, while the history of science is of peripheral interest to a modern scientist, 'The History of Philosophy is and always has been part of philosophy; and philosophy is much more concerned with its past than any other academic discipline.'[18] The contrast has been taken further by Richard Rorty. Historians of science, he notes, 'do not think it anachronistic to say that Aristotle had a false model of the heavens, or that Galen did not understand how the circulatory system worked. We take the pardonable ignorance of great dead scientists for granted.' Historians of philosophy, however, are less willing 'to say that Aristotle was unfortunately ignorant that there are no such things as real presences, or Leibniz that God does not exist, or Descartes that the mind is just the central nervous system under an alternative description'. The reason for such hesitation, Rorty concludes, is merely because:

we have colleagues who are themselves ignorant of such facts, and whom we courteously describe not as 'ignorant', but as 'holding different philosophical views'. Historians of science have no colleagues who believe in crystalline spheres, or who doubt Harvey's account of circulation, and they are thus free from such constraints.[19]

The contrast has been further marked, and in a more objective fashion, by Derek Price. He surveyed 154 scholarly journals to find out what percentage of references dated from the previous five years. The highest figure came from such scientific journals as *Physical Review*, in which 73 per cent of works cited had been published in the previous five years. Comparable figures for the philosophical journals *Mind, American Philosophical Quarterly*, and *Philosophy of Science* were 24, 18, and 21 per cent respectively.[20]

As with earlier distinctions, the contrast is too sharply and selectively marked. There is no difficulty in referring to philosophers who do espouse ancient doctrines. A. A. Luce, for example, not only spent much of his life editing the works of Berkeley, he also, in his *Berkeley's Immaterialism* (1945), offered a passionate defence of the views Berkeley had put forward almost 250 years before. Equally, it is not difficult to find scientists who are scarcely willing to concede that their subject even has a history.

These are, however, extreme cases, untypical of anything more significant than the variety of human tastes and prejudices. A more balanced account would acknowledge that many scientists are deeply aware of and often profoundly knowledgeable of the historical antecedents of their work. Moreover, they frequently go out of their way to place their own work in some kind of historical context. However formidable the text, however textbook-like it might appear, historical considerations are likely to find some place.

The reason for this is clear. Scientists are well aware how vulnerable their work is likely to be to the critical scrutiny of their successors. Anyone who presents his views as timeless truths is very likely to sink into oblivion when they are overthrown by a later generation. Once historicised, however, and whatever their fate, their views will be preserved forever as part of a still flourishing tradition.

But the complaint is not of awareness of the past, but of subservience to it. Rorty has philosophical colleagues who believe in a Leibnizian God and a Cartesian mind; none of his scientific colleagues subscribe, however, to Galen's physiology or Ptolemy's astronomy. Rorty's choice of example is most curious. For the first believers in Leibniz's God and Cartesian souls who spring to mind are scientists, not philosophers. It is, for example, the neurophysiologist and Nobel laureate, Sir John Eccles, who has argued most recently and most strongly for Cartesian dualism.[21] Again, supporters of the view that the universe is endowed with reason are most likely to be found amongst the cosmologists and theoretical physicists who have sought to establish some form of anthropic principle.[22] Most notably, however, Cartesian ideas on the nature of mind and language have been influential in the development of the work of the linguist Noam Chomsky. The nature of this influence has been described by Chomsky himself in his *Cartesian Linguistics* (New York, 1966).

Nor are scientists quite so free from or indifferent to the influence of their own past. Success in science of a substantial kind frequently brings with it much more than a number of new and important truths. A particular method, style, and approach to nature will often prove to be more influential than the actual results established by their exercise. Thus, Newton's legacy to later generations of scientists consisted of much more than his laws of motion. The Newtonian method with its positive commandments and its accompanying prohibitions, as will be seen below, served as a model for many later generations of scientists.[23]

Also of interest is the evidence of the textbooks. They are not always exclusively modern, or invariably indifferent to the work of their distant predecessors. A once widely used work in the earlier part of the century was *A Text-Book of Physics* in four volumes by J. H.

Poynting and J. J. Thomson. It is instructive to see how they deal in the volume on *Heat* (1904) with a fairly straightforward issue like the subject of chapter 2: 'The Expansion of Solids with Rise of Temperature'. After some initial definitions they move straight into a description of Ramsden's method 'devised in 1785'. This is followed by an account of the 'Method of Lavoisier and Laplace', also dating from the 1780s, and thereafter of Fizeau's method dating from the 1850s. Lengthy quotations are given from original sources, and obscure points of ancestry are pursued. The approach is followed throughout the four volumes with judgements on past science being made which are neither reverential nor patronising. There is, instead, a simple awareness that many issues and problems are best understood as part of a still flourishing tradition. And just as students of philosophy approach the problem of causation through the writings of Hume, Kant, and Mill, readers of Poynting and Thomson will find themselves dealing with the determination of the gravitational constant through the eighteenth-century experiments of Bouguer, Cavendish, and Maskelyne.

In contrast to this approach some philosophers have consciously resolved to strike out anew, with as few debts from the past as possible. They do not wish to discount entirely the work of past philosophers; they merely emphasise that the debates of the past have brought with them more confusion than clarity. For this reason the Oxford philosopher J. L. Austin recommended the subject of 'excuses' for study, precisely because it was not 'too much trodden into bogs or tracks by traditional philosophy'. The concepts 'clumsiness' and 'inconsiderateness' were attractive to Austin just because they could be discussed 'without remembering what Kant thought'. Within the subject of 'excuses' we could even hope to discuss deliberation 'without for once remembering Aristotle', or 'self-control without Plato'.[24]

So far nothing has emerged which could be considered in any way conclusive or definitive. There remains, however, one further criterion which has gathered a considerable following. Scientists solve problems; philosophers, along with historians and other humanists, it is argued, only have opinions. The issue is sufficiently complex to require a chapter of its own.

3

IS SCIENCE REALLY THE ART OF
THE SOLUBLE?

> If politics is the art of the possible, research is surely the art of the
> soluble.
>
> Peter Medawar, *The Art of the Soluble* (1969)

Science is pre-eminently concerned with solutions. Not any old solutions, of course, but exclusively with those which are definitive, permanent, and which command wide support. The point is a familiar one and was made, for example, by Norman Campbell in his influential *What is Science?* (1921). Science, Campbell argued, was 'the study of those judgements concerning which universal agreement can be obtained'. The point was repeated half a century later by John Ziman in his equally influential *Public Knowledge*: 'The objective of science is not just to acquire information nor to utter all non-contradictory notions: its goal is a *consensus* of rational opinion over the widest possible field.'[1] The view, however, has received its most memorable expression in Peter Medawar's oft misquoted epigram: 'If politics is the art of the possible, research is surely the art of the soluble.'[2]

Nor is the position implausible. Give a scientist a problem and he will probably provide a solution; historians and sociologists, by contrast, can offer only opinions. Ask a dozen chemists the composition of an organic compound such as methane, and within a short time all twelve will have come up with the same solution of CH_4. Ask, however, a dozen economists or sociologists to provide policies to reduce unemployment or the level of crime and twelve widely differing opinions are likely to be offered. Medawar's epigram seems to recognise this point and to locate it in its rightful position as the central feature distinguishing science from all other intellectual disciplines, including, of course, philosophy.

The point has been made, at least in part, often enough by philosophers ever since Socrates first complained about the lack of agreement between the various Athenian schools. Philosophers have also, at regular intervals, exhorted their colleagues to pull themselves together, adopt a little self-discipline, and apply themselves more seriously to the task of establishing some agreed solutions to some of philosophy's outstanding, unsolved problems. Thus, Ayer in his *Language, Truth and Logic* (1936) complained strongly about the party divisions so typical of philosophy. He insisted also that at last he knew how such long-standing rivalry could be dispelled.

Perhaps it was the interruption of the war years which delayed the implementation of Ayer's radical programme. Whatever the reason, similar sentiments were being expressed by Gilbert Ryle in 1945 on his return to Oxford from service with the Brigade of Guards. Sufficient time had been spent discussing methodology and the nature of philosophy, he argued, and the time had come to solve a philosophical problem:

It was time, I thought, to exhibit a sustained piece of analytical hatchet-work being directed upon some notorious and large sized Gordian knot. After a long spell of enlightened methodological talk, what was needed now was an example of the method really working, in breadth and depth where it was really needed.[3]

For a time he considered tackling the problem of the freedom of the will, but in the end 'opted for the Concept of Mind'.

Yet Ryle's classic work of the same title, published in 1949, fared no better than Ayer's 1936 volume. A decade later another Oxford philosopher, J. L. Austin, could be found complaining once more that philosophy had been barren for far too long, that it lacked 'the fun of discovery . . . and the satisfaction of reaching agreement'.[4] Like Ayer and Ryle before him, Austin also had a method. Its deployment, he expected, would lead to the discoveries and consensus so frequently met with in science, and so conspicuously absent from philosophy. Austin's optimism turned out to be as misplaced as his predecessors'.

Is, therefore, Medawar's thesis as sound as its initial plausibility might suggest? Does it, in fact, mark a crucial distinction between science and philosophy? No direct or simple answer can be offered to this apparently straightforward question. The thesis is in fact a complex one and can be broken down into the following four claims:

1 Science, unlike philosophy, aims to solve problems.
2 What is more, science, unlike philosophy, often succeeds in solving its problems.
3 Further, such solutions, unlike philosophical solutions, are not merely partisan; they command wide support.
4 Finally, scientific solutions are true solutions.

The issues expressed in the above claims will be discussed further under the separate headings *Solutions*, *Success*, and *Consensus*. The fourth issue, *Truth*, will be considered at greater length in chapter 11 below.

Solutions

Surely, it must be obvious to all that scientists aim to solve the problems of science. The better they are as problem-solvers, the better they are as scientists. Within this context few scientists are as good as Francis Crick, the molecular biologist. A colleague, the biochemist Roger Kornberg, spoke revealingly of him as quite 'ruthless' in his determination to solve problems: 'He would just carve them up and solve them in the most brutal way.'[5] Fermi had a similar reputation among physicists.[6]

Yet, surprisingly, Kornberg seems to be suggesting that Crick was a far from typical scientist in this respect. Erwin Chargaff, for example, another biochemist, was dismayed by Crick's approach.[7] Kornberg went on to claim that most scientists have a tendency to 'fondle' their problems. They end up eventually working on problems they cannot bear to see solved. At this point their lives as creative scientists end, a fate awaiting all aged scientists. Many have claimed to detect such a state of mind in Einstein's later work.

Hoffmann has described how in the 1920s Einstein, dissatisfied with quantum theory, began his search for a unified field theory. He was not alone. For a time:

> The search for a unified field theory became a mathematical fad indulged in by many people, famous and not, who produced an enormous number of competing geometrical theories. When the fad began to die down, Einstein continued. But still he could find no physical guidance, no magical insight, and because of this, many physicists looked upon his long search with barely concealed contempt.[8]

At his death in 1955 Einstein was still working on the problem. He was well aware that his thirty-year commitment to an unsolved problem appeared strange to many of his colleagues. In defence,

> He agreed that the chance of success was very small, but that the attempt must be made. He himself had established his name ... he could afford to take the risk of failure. A young man with his way to make in the world could not afford to take a risk by which he might lose a great career, and so Einstein felt that in this matter he had a duty.[9]

Alongside the great problem-solvers of the day, we can recognise also a large number of scientists, less well known perhaps, who seem more concerned with other issues. The name of Johann Schmidt (1825–84) is seldom met with in accounts of the nature of science. Schmidt spent much of his life studying the moon's topography. The results of his labours were presented in his *Charte der Gebirge des Mondes* (1878) on a lunar map with a diameter of 195 cm. Among other things it recorded some 30,000 craters. A contemporary, Friedrich Argelander (1799–1875) in his *Bonner Dorchmusterung* mapped the approximate position of 324,198 stars. The mathematician Charles Delaunay (1816–72) began in 1847 to calculate the position of the moon as a function of time. Previously astronomers, because of the complex nature of the lunar orbit, had tended to work out the moon's position by extrapolating from its earlier positions. Delaunay's calculation took twenty years and was finally published in 1867. When checked by computer in 1970 the calculation took twenty hours and revealed three small errors. Finally, we may consider the case of R. E. Reason, an engineer who devoted his life to developing techniques for measuring roundness. He succeeded in attaining accuracies better than 100 nanometres.

So scientists, as well as solving problems, assemble data, catalogue, classify, calculate, tabulate, measure, describe, assess, and, among many other activities, collect. Work of this kind is as much part of science as the accounts of the origin of the universe or of the nature of gravity offered by better known and more theoretically inclined scientists. It could even be argued that in many ways they are more durable and more central features of science than the search for solutions. Delaunay's methods, for example, suitably modified, are used today in satellite tracking. Even the stellar catalogues of Hipparchus dating from the second century BC are of some interest to the astronomers of today. In contrast the solutions Hipparchus offered to the problem of the retrograde movements of the planets concern only historians of ancient astronomy.

The assumption seems to be that without agreed solutions there can be no progress. This is far from true, even of science itself. One rich source of unsolved problems is number theory. Probably the most famous unsolved problem of number theory was first posed in 1665 and is known somewhat ironically to mathematicians as Fermat's last theorem.

Students of geometry are familiar with the equation $x^2 + y^2 = z^2$, and how it is satisfied by, amongst others, the Pythagorean numbers 3, 4, 5. Can the equation be generalised for any higher exponents? Are there any numbers $n > 2$ such that $x^3 + y^3 = z^3$? A clear answer was found after his death in 1665 in the margins of Fermat's own copy of Diophantus. He had discovered, he noted, 'a truly marvellous demonstration' of the proposition that the equation $x^n + y^n = z^n$ was not satisfied by any $n > 2$. Unfortunately, Fermat concluded, the margin was too narrow to contain his proof!

Three centuries later, despite the offer of large financial rewards, mathematicians are still seeking a proof. Their failure to discover a proof, or a disproof, does not mean that no progress has been made. Fermat himself was able to prove that the equation does not hold for $n = 4$, and in the following century Euler did the same for $n = 3$. By 1859 Kummer was able to show that the equation did not hold for any $n < 100$. Since then high-speed computers have extended the result to $n = 125,000$. Clearly, if there are numbers x, y, z, which do satisfy the equation they must be very large. In addition, Kummer's work on the problem led him to introduce into mathematics a new class of objects, ideal numbers, and to develop with them a theory of ideal factorisation which has been described as 'one of the major achievements of 19th-century mathematics'.[10]

The example is typical of work on many other problems in both science and philosophy. Some progress is made, and interesting new ideas developed. Consider, however, the notions of solutions and progress as they can be applied to the game of chess. In many ways chess in this respect is very like philosophy. Both have long histories which most practitioners are aware of and frequently consult. The chess master is just as likely to be found replaying the past games of Morphy and Capablanca as the modern philosopher is to be found rehearsing the arguments of Descartes and Kant. Further, chess games, like philosophical problems, can often be sufficiently complex to defy complete analysis. It has been calculated that after only two full moves $71,852$ legal positions can be reached, while the number of distinct forty-move games is the colossal figure of 25×10^{115}.

Chess is also similar to philosophy in possessing a number of long-standing unsolved problems. There is for white, however expert, no opening which will guarantee victory, and no response from black which can be relied upon to preserve him from defeat. Nor is it absolutely clear which is the best response to many openings. Some masters prefer one reply, others favour a different response.

Despite these shortcomings, it would be wrong to conclude that no progress has been made in chess, or that it is an activity without method or structure. Chess masters possess skills other than the ability to win. They are also extremely good at analysing games and positions. And these are skills which have improved over time and benefit from experience. Asked about an opening strategy, or a particular position, the master will relate it to his own experience and the numerous games of other masters. In addition to his own formidable chess insight, he can call on the experience of his contemporaries and predecessors. He might note that in a similar position Alekhine had found it difficult to develop his queen-side pawns, or that intense analysis favoured a rook sacrifice at a certain point. These are objective judgements based on many past games, and they carry with them considerable predictive accuracy. That is, players who ignore the warnings of the past, and the analysis of experts, are indeed likely to find themselves, as warned, unable to develop their queen-side pawns.

Whether or not this turns out to be a fatal defect will inevitably depend on the relative skills of the players. Any positional weakness can be rescued by an opponent's errors.[11] Only too aware of the unpredictable outcome of most chess *games*, the editors of *Modern Chess Openings* restrict their attention to *positions*. And here, of course, they are invariably incapable of offering a complete analysis. Consequently, they annotate the various openings and positions with a useful set of signs designed to indicate the probable outcome of a particular strategy. Thus '++' indicates 'a clear winning advantage', '+' a 'distinct superiority', and '±' and '∓' stand for 'slightly better chances' for white and black respectively. Less favourable are '?' to indicate 'a weak move', and '!?' to show a merely 'speculative attempt to complicate'.[12]

The signs could be used just as plausibly to annotate many a philosophical discussion. In the manner of the chess master the philosopher is seldom in a position to offer definitive solutions to major problems. But, also like the chess master, while he may be unable to offer exhaustive analyses of the mind–body problem, the freedom of the will, and the problem of knowledge, the philosopher will be very much aware of the strengths and weaknesses of various approaches to these problems. He will know where lines of argument lead, and the difficulties they will eventually have to face. And, while he may be acutely aware of his own inability to advance any of the issues further himself, he may confidently feel that he can annotate any issue in the manner of *Modern Chess Openings*. And where the chess master will justify his judgements by appealing to a move of Capablanca in 1925, the philosopher will appeal to a distinction drawn by Moore or Russell in *Mind* some time ago. An example of this kind of approach can be seen in the treatment given by David Armstrong to a traditional difficulty in the theory of knowledge.[13]

At one point, for example, he considers the problem of 'the infinite regress of reasons'. It has been held that knowledge is 'true belief for which the believer has sufficient evidence'. But what of this evidence q? For it to be sufficient grounds for my claim to know p, I must first know q. But this leads to an infinite regress in that before I can know p, I must first have knowledge of some evidence q, which in turn requires me to have established my knowledge of q on some earlier evidence q' which, of course, I must also have known. And so the regress continues. It was first recognised by Plato in the *Theaetetus* (209e–210b). In the manner of *Modern Chess Openings* Armstrong lists six 'different reactions to the threatened regress'. Briefly, they consist of the following:

1   The sceptical reaction which, in the manner of Popper, accepts the regress and concludes that we should prefer hypotheses to knowledge.

2   The regress is indeed infinite but not vicious. [Armstrong is only prepared to accept 1 or 2 if all other responses fail.]

3   The regress is finite and coherent: p is supported by q, which is supported by r, which is supported by s, which is supported . . . by p; we have 'a circle of true beliefs which mutually support each other'. This depends on an account of truth earlier rejected by Armstrong.

4   The regress, in the manner of Descartes, ends in self-evident truths. Armstrong denies the existence of such truths.

5   The regress ends, not in the self-evident propositions of (4), but in 'initial credibility'.

6   Externalist theories in which some natural relation holds between the belief state and the situation which makes the belief true. They are subdivided into causal and reliability theories. [How can knowledge of the future be caused by future non-existent events?]

Armstrong himself favours some form of reliability theory, a view he traces to Plato's *Meno* (97b).

Science also has a number of long-standing problems lacking solutions. But the fact that mathematicians can neither prove nor disprove Fermat's last theorem does not mean they lack all insight into the problem. Three hundred years' work has, at least, established where some difficulties lie, and where cul-de-sacs abound. A mathematician choosing to work on the problem without consulting the literature could spend his whole life merely rediscovering the errors of his predecessors.

There is one final figure, seldom given prominence in accounts of science, who constructs no theories, and seeks no solutions. He is the critic and he seeks to destroy the solutions of his colleagues, to expose their weaknesses and difficulties, rather than to offer further solutions of his own. His role need not always be sterile or minor. Acute and perceptive criticism, though it may well lead nowhere, can at least expose certain lines of research as unpromising.[14]

For example, students of Darwin have long been familiar with the crucial role played by Fleeming Jenkin, a professor of engineering at Glasgow University, in the development of evolutionary theory. In 1867 Jenkin published an influential review of Darwin's *Origin of Species* in which he presented a number of very general arguments against Darwin's account of natural selection. It was widely assumed that inheritance was a form of blending, that parental features were to be found suitably mixed in their offspring. Darwin also assumed that natural selection operated on the individual variation found present in any population. Evolution, Jenkin pointed out, simply could not work on these assumptions. Even though certain variations could well have survival value, Jenkin argued, their presence would be rapidly diluted. Thus, he pointed out, in the racial language of his day, a white man wrecked on an island of blacks, though he may end up as king, 'cannot blanch a nation of negroes'.[15] With this contribution Jenkin's role in the history of biology ended. Biologists, however, were forced to struggle for the rest of the century to reconcile the theories of evolution and inheritance.

Jenkin had hit upon a genuine difficulty. Scientific criticism is not always so well based. Consider, for example, the case of Max Von Pettenkofer, one of the first to emphasise the role of hygiene in medicine and, with his *Handbuch der Hygiene* (1882), the author of one of the first scientific textbooks on the subject. Surprisingly, such a figure, a contemporary of Pasteur and Koch, spent much of his life battling against the germ theory of disease. In one dramatic gesture he drank a sample of cholera germs and subsequently cabled Koch the news that he remained in good health.[16]

Von Pettenkofer had a number of far from trivial arguments to support his position. How could germs spread disease, he asked, when doctors, nurses, and other health workers seemed to be no more prone to infection than the population at large? Again, how could a disease like cholera suddenly appear in an *isolated* community? Or, why did Paris and Marseilles succumb to cholera epidemics while Lyons and Wurzburg reported no infections? Did no one, Von Pettenkofer scathingly enquired, ever travel from Paris to Lyons?[17]

In more recent times the role of Von Pettenkofer has been frequently re-enacted on a much wider front by the astronomer Fred Hoyle and his collaborator, Chandra Wickramasinghe. While the rest of the scientific world was to be found in 1983 celebrating the centenary of Darwin's death, Hoyle could be found repeatedly insisting that natural selection was a most implausible hypothesis. *Archaeopteryx*, the most famous fossil of all, the pride of the Natural History Museum, and the most important evidence for the evolution of the birds from the reptiles, was dismissed simply as a 'fossil forgery'. Nor is Hoyle much impressed with the modern germ theory. Diseases such as bubonic plague and influenza could have fallen from the air rather than having been transmitted through the migration of rats, or by coughs and sneezes. The evidence is mainly epidemiological. No plague apparently hit Bohemia and southern Poland, although both areas 'grow food just as palatable to rats as everywhere else'. Or, to turn to flu, Hoyle and Wickramasinghe pointed out that in 1918 the virus could spread among the widely scattered 45,000 inhabitants of Alaska despite the fact that 'human travel from the coast to the interior was essentially impossible because of snow and ice.'[18]

Such examples do little to convey the richness, precision, and range of Hoyle and Wickramasinghe's arguments. Characteristically these are somewhat unorthodox, and they come from scientific trespassers. Hoyle is no palaeontologist, virologist, or bacteriologist, and perhaps because of this he can more readily tolerate the collapse of orthodox theory. Whether the points raised by Hoyle and Wickramasinghe are sound remains to be seen. Radical critics are seldom entirely correct in their claims. Their importance lies more in the challenge they offer than in the propositions they establish.

There are, of course, large numbers of critics, ever present, and ever ready to challenge any theory. Their objections, however, are often technical, and argued from within the orthodoxy. They seldom challenge the entire theoretical structure, and more often seek to do no more than draw attention to problems and anomalies. They seek basically to improve, and not to overthrow. The role of the radical critic is quite different. He asks questions others would never dream of raising. He challenges what is normally taken to be most secure, and queries the deepest of assumptions. Because of his existence these principles can be resurveyed and, if found satisfactory, as they invariably are, can be held even more firmly. The role of the radical critic, whether he be Hoyle, Velikovsky, or Sheldrake, more often serves to reinforce initial assumptions than to overthrow them. In precisely the same way conservative politicians never feel so sure of their beliefs as when they hear their cherished principles savagely rejected by their socialist rivals.

Success

The nineteenth century saw the emergence of the view that science is by far the most systematically successful intellectual discipline ever to emerge. Magic, theology, philosophy, or whatever, have all had their day. None, however, have succeeded, to the satisfaction of so many, on so many issues, over so many fields. One of the first to preach the new gospel was Macaulay. In 1837, in a famous essay on Bacon, he spoke of how science had:

lengthened life . . . mitigated pain . . . extinguished diseases . . . annihilated distance. These are but part of its fruits, and of its first fruits; for it is a philosophy which never rests . . . Its law is progress. A point which yesterday was invisible is its goal today, and will be its starting-point tomorrow.[19]

Compare this to the attitude of Seneca when forced to face the awful prospect of having to admit that the philosophers Anarchasis and Democritus had invented the potter's wheel and the arch respectively. Perhaps, Seneca conceded, but not *as* philosophers. They were like an athletic philosopher who wins a race, 'by virtue of his ability as a runner, not by virtue of his being a philosopher'.[20]

Nor was the success of science seen exclusively in such utilitarian terms. Writing of the 1870s and 1880s, Beatrice Webb recalled that it was 'the men of science' who were the leading British intellectuals of her youth: 'it was they who were routing the theologians, confounding the mystics, imposing their theories on philosophers, their inventions on capitalists, and their discoveries on medical men.'[21] At the same time the historian of philosophy, George Lewes, was writing dismissively of philosophers wandering 'for ever in one tortuous labyrinth' wherein they are 'continually finding themselves in the trodden tracks of predecessors who could find no exit'.

Whatever has changed, the emphasis on success has remained constant. It does, however, tend to be expressed in a somewhat different idiom. Quantum theory, we are now told, is the most successful branch of science. So successful, John Bell insists, that it is difficult to believe that it could be wrong. Nick Herbert has found it to be 'flawlessly successful at all levels accessible to measurement'.[22] And the much respected theoretician John Wheeler has claimed it to be 'unshakeable, unchallengeable, undefeatable – it's battle tested'.[23] The basis for such extravagant claims lies in the manner in which extremely accurate predictions, unsuspected in classical physics, have been derived from quantum mechanics and triumphantly confirmed.

Consider, for example, the electron. As it is a charged particle, and as it spins, it will act like a small magnet. It will therefore possess a magnetic moment. If only classical considerations are allowed, the electron's magnetic moment (g) takes the value 1. These considerations, however, fail to take into account a number of familiar properties of the electron including the Zeeman and the Stern-Gerlach effects. Classically g could be identified with $e/2m$ where e is the electron's charge and m its mass. In quantum mechanics, however, g is measured in units of $eh/4\pi m$, where h is Planck's constant, which on calculation turns out to be 2.0023193048. Experimental measurements of g, almost unbelievably, produced the same 2.0023193048.[24] Conflict between theoretical calculation and experimental measurement appeared only with the eleventh decimal point. Not surprisingly physicists speak of the 'astonishing accuracy' of their work.

It would, of course, be absurd to deny that at one level science is enormously successful. Around us we have the synthesised steroids, lasers, and microchips, as daily reminders. Yet a number of qualifications are called for.

Is it perhaps the case that scientists thrive as problem-solvers only in certain restricted areas? Give a palaeontologist a fossil, or a sample to an analytical chemist, and they will very probably be able to identify the objects. What, though, of other problems? Medawar's aphorism appeared in response to the complaint made by Arthur Koestler that scientists ignore a number of important problems. Why, for example, had they so little to say about 'the genetics of behaviour'? Medawar conceded the point and went on to explain:

Goodness knows how it is to be got at. It may be outflanked or it may yield to attrition, but probably not to direct assault. No scientist is admired for failing in the attempt to solve problems that lie beyond his competence. The most he can hope for is the kindly contempt earned by the Utopian politicians.[25]

Science thus guarantees its success by carefully restricting its attention to just those problems it knows how to solve. Medawar's aphorism preserves its truth at the cost of triviality. In this form the point had been made long ago by William James in his declaration that philosophy was the collective name for 'questions that have not yet been answered'. Whenever solutions appear on the distant horizon, James argued, scientists make an appearance and take the credit, while unanswered residua remain 'to constitute the domain of philosophy'.

Further, if the focus is directed away from the topical successes of quantum theory to less tractable areas, the successes of science are less apparent. The great German mathematician David Hilbert was once asked what question he would first ask if he returned to earth in five hundred years' time. He responded without hesitation: 'Has someone proved the Riemann hypothesis?'[26] Hilbert's response hardly suggests that he had much confidence in the mighty power of modern mathematics to solve one of its deeper problems. Again, many sufferers from migraine, cancer, arthritis, virtually any virus disease, or Alzheimer's syndrome, may also be somewhat sceptical about the power of science to solve their particular problems.

It must also be accepted that yesterday's scientific solutions often turn out to be the errors of today. Choose the right problem, the size of the universe for example, and the history of science can be seen as no more than the replacement of one error by another. By way of illustration, consider the estimates of this measurement made by astronomers from the time of Ptolemy onwards, listed overleaf. Figures are expressed in earth radii.[27]

Surveying this record is hardly likely to convince the general reader that today's solutions will fare better than their predecessors. Indeed, on this and a number of similar topics, successful solutions

Ptolemy (*c.* AD 150)	20,000
Al Farghani (9th cent.)	20,110
Al Battani (9th cent.)	19,000
Al Urdi (13th cent.)	140,000
Levi ben Gerson (14th cent.)	159,651 billion
Copernicus (1543)	7,850,000
Brahe (1602)	14,000
Kepler (1609)	34,177,000
Kepler (1619)	60,000,000
Galileo (1632)	2,160
Riccioli (1651)	200,000
Huyghens (1698)	660,000,000
Newton (1728)	20 billion
Herschel (1785)	10,000 billion
Shapley (1920)	1,000,000 billion
Current value	100,000 billion

are likely to be rare. How large is the universe? How old? How dense? How did it begin? How, if at all, will it end?

It might be thought that on such issues as the size of the universe scientists could be expected to speak tentatively, and that it is the very nature of such judgements that they be revised regularly. On other and more prosaic issues, however, solutions could be offered with greater confidence. The question of how much matter there is in the universe is clearly very different from questions about the melting point of iron. Yet, even with very limited and precise questions of this kind, science can often speak for a century or more with no coherent voice.

Consider, for example, the seemingly simple case of the giant panda, first described by a western observer in 1869 when Père David obtained a skin from western Szechwan. What kind of creature was it? David took it to be a bear and therefore named it Ursa melanoleucus. A specimen was sent to Alphonse Milne-Edwards at the Paris Natural History Museum. He considered it to be closer to the raccoons and consequently renamed his specimen Ailuropus melanoleucus. Thereafter, for a century or more, experts have continued to disagree on the status of the giant panda. George Gaylord Simpson in his influential *Classification of the Mammals* (New

York, 1945) placed the giant panda with the raccoons, while D. D. Davis in his equally influential and authoritative monograph, *The Giant Panda* (Chicago, Ill., 1964), declared it to be a bear.

Nor was the issue simply a choice between being a raccoon or a bear. If the giant panda was thought to be a bear, the question of its relations to other bears immediately arose. Did it form, as Winge argued in 1895, a distinct group whose only close relative is the extinct Hyenarctos? Or is it more closely related to living bears? Similar problems arise if the raccoon option is adopted.

But, in a recent study, S. J. O'Brien, despite conceding that 'for almost 120 years biologists have disagreed about the correct taxonomic placement of this species', now feels free to claim that 'the riddle of the panda has now been solved.' Beginning in the 1970s, more exact biochemical techniques began to be applied to the problem. On the basis of immunological tests V. Sarich published a paper in *Nature* in 1973 with the title 'The Giant Panda is a Bear'. Further work described by O'Brien on the DNA and chromosomes of the pandas, bears, and raccoons apparently supports Sarich's claim. At last zoologists can speak on the issue with some authority and coherence, but it remains to be seen how durable the present consensus really is.[28]

It is also necessary to point out that the above account of the search for the taxonomy of the giant panda is fairly typical of normal science. In any one year *Scientific American* and similar journals will carry many articles of this kind. After reviewing the disputes of the previous century, half-century, or whatever, the author will invariably conclude by noting that, with more exact modern techniques, the issue can now be decisively resolved. Once more, it remains to be seen whether such confidence is misplaced.

On a more pedestrian level it has been pointed out that many of the supposed solutions of modern science are in fact illusory. An estimate by R. Roberts in 1977 concluded that 'at least half of all published scientific papers were unusable or unreliable.' Even in the field of mathematics, Zahler's study has shown, 'many published articles undoubtedly contain serious errors.'[29] Nor need this be entirely a matter of incompetence. Some scientists are simply attracted to the more difficult problems of their day. One such unnamed physicist, for example, was described by Einstein in a conversation with Phillip Frank. The physicist, Einstein noted, had gained little success:

Mostly he attacked problems which offered tremendous difficulties. He applied penetrating analysis and succeeded only in discovering more and more difficulties. By most of his colleagues he was not rated very highly. Einstein, however, said about him, 'I admire this type of man. I have little patience with scientists who take a board of wood, look for its thinnest part and drill a great number of holes where drilling is easy.'[30]

Consensus

As has been seen, scientists are prone to characterise their discipline in terms of its ability to achieve a wide consensus. The same point has been picked up by philosophers; for example, David Armstrong claims that 'It is only as a result of scientific investigation that we ever seem to reach an intellectual consensus about controversial matters.'[31] By contrast, all other disciplines – history, philosophy, sociology, and theology – are assumed to be perennially divisive.

The picture as presented is clearly a caricature of both science and the other disciplines. What, it must be asked, is the source of this remarkable consensus? There are many forms of consensus. There is, for example, the consensus of fashion, through which almost overnight millions of individuals decide that nothing could be worse than to be seen in public in flared trousers. There is also charismatic consensus, whereby a powerful leader, religious or secular, imposes his beliefs upon his followers. More significantly, and more relevantly, there is the consensus which arises from the procedures of a discipline.

We can all agree that 43.29×17.85 equals 772.7265 because there is a mathematical procedure we can all follow which, when correctly followed, invariably leads to the same answer. Even if we read the answer directly from a calculator we cannot escape the appropriate procedure; we merely transfer it to the electronic level. Any consensus reached is simply a consequence of the existence of such algorithmic procedures. If a wider consensus is to be found in science, in physics and chemistry as well as mathematics, then comparable procedures must be spread widely throughout science.

Present a molecule to a group of chemists, or a fossil to a palaeonto-logist, and, as with the calculation, they are likely to follow common procedures leading them to agreement on the structure of the molecule and the age and nature of the fossil.

The difficulty with this idea of consensus is that it fails to identify science. Consensus in this sense can be reached as readily in literature, history, and philosophy, as in chemistry and geology. There are, for example, procedures to be followed in publishing such important texts as the letters of Erasmus and the diaries of Pepys. In brief, the aim must be to publish as full and accurate a text as possible while at the same time providing whatever background is necessary to make the text intelligible. Consequently, it is widely accepted that Braybrooke's edition of Pepys published in 1825 is inadequate. It is described by Pepys's modern editor, Robert Latham, as a 'travesty'.[32] When it is realised that Braybrooke omitted without comment whole passages, that words and phrases were added, also without comment, and that his notes were 'few, brief, and not always correct', it becomes clear why modern scholars dismiss his work. And, also, why there is a similar but favourable consensus about Latham's own edition of Pepys, as there is about the works of Locke, Pope, Horace Walpole, or Dryden as they can be read in their various definitive editions. Such works may well be the result of a century or more of effort but, because of it, we can now accept with confidence that the words of Gladstone, Washington, or Voltaire are accurately presented in the massive editions of their collected correspondence.

It is of course true that more contentious issues can be identified without difficulty. The significance and value of the novels of D. H. Lawrence, for example, or the number of African slaves shipped across the Atlantic, are likely to be as contentious as any issue in politics. But, given the right issue, similar diverse responses can be met with in science. Larry Laudan, for example, has made the sensible point that both consensus and dissensus are to be met with in science, and that it is the task of any adequate philosophy of science to be able to cope with both features. He refers to the 'ubiquity of controversy' and notes how the history of science, no less than literature, has been characterised by persistent, genuine disputes between its most important practitioners.[33] None more so, perhaps, than the dispute over the nature of light which, beginning with

Newton and Huyghens in the seventeenth century, was still being actively pursued in the mid-nineteenth century. Newton argued that light had a corpuscular character. He dismissed the claims of Huyghens that light moved with a wave-like motion on the ground that waves 'bend manifestly' whereas light moves rectilinearly. Huyghens and his supporters objected with equal plausibility that light beams made of particles would be unable to interact without destroying each other.

These and other arguments were repeated endlessly throughout the eighteenth century. On this issue, Laudan notes, 'the rules and norms of science' were clearly insufficient to resolve the dispute. Several other prolonged disputes have been mentioned and range in time from the Copernicus–Ptolemy debate to the Bohr–Einstein confrontation on quantum mechanics. So prevalent are they that they have seemed to some commentators not as an unfortunate aberration of science, but as an essential part of its dynamics. Thus, Thomas Kuhn has claimed that the ability of science to generate new theories actually '*requires* a decision process which permits rational men to disagree'. If, he went on to argue, some philosophers were right in their view of science, then we would find 'all conforming scientists would make the same decision at the same time'.

So far it has been argued that science is not quite so successful, or so marked for its consensus, as some scientists have been prone to claim. What then of philosophy? Must we accept the oft-repeated charge of Descartes against the philosophy of his day, that there is 'no single thing . . . which is not the subject of dispute'? And must we also accept the comment of Kekes: 'There has been no improvement since this passage was written'?[34]

Some would side with William James in his claim that as philosophy is a name 'for questions that have not been answered' it is unsurprising that philosophy is not awash with either solutions or consensus. As soon as the prospect of a solution appears on the horizon, it is likely to be hijacked by scientists as their own problem. The answer is too glib and concedes too much to our critics. It might explain an occasional development in philosophy; it throws no light however on its nature. Much of philosophy, of course, is not about solutions at all. At a very simple level it is a criticism of other people's solutions, and when it does venture into the business of solutions it does so self-consciously and with no real commitment.

To say it is essentially a critical approach to others' solutions could well make the subject appear unworthy, and charges that it is un-creative, parasitic, carping, and pointless, are likely to be levelled fairly quickly. In refuting this formidable attack, one further feature, suggested by the mathematician G. H. Hardy, may be relevant. None of his own kind of pure mathematics, he noted, was 'directly useful' in the manner of chemistry and physiology. What then was its value? One point that struck him was the character of permanence that went with much mathematics. And, he emphasised, for anyone to have produced anything of the slightest permanent interest was no mean thing.[35] A similar property can sometimes be found within philosophy.

Where then are these permanent features of philosophy to be found? Not, perhaps, in the multi-volume tomes of speculative philosophy. They have the permanence of print, but are little read and long out of print; they lack the permanence of continuous use. Durability of this kind, however, is to be found more frequently among the less ambitious principles, distinctions, arguments, analyses, paradoxes, fallacies, and theorems propounded by philosophers over the centuries. Some of them, like the Cartesian *cogito*, Ockham's razor, Kuhn's paradigm, and the positivist's verification principle have indeed become part of our general intellectual background. So much so that a well-known computer firm can advertise its products with the slogan 'I think therefore IBM.'

A good philosophical argument is not necessarily a valid one. It is, rather, an argument of sufficient depth and width to guarantee its independence, durability, and suggestiveness. That is, it cannot readily be reduced to any simpler argument, defies instant analysis, and often leads to other interesting positions. In this sense such arguments as Aristotle's sea-battle (see below) and Zeno's Achilles and the Tortoise are still with us.

Another sign of a good argument is that philosophers find themselves bumping into it unexpectedly. One example will suffice. In chapter 9 of his *De Interpretatione* Aristotle turned to the problem of future contingents. The proposition 'There will be a sea-battle tomorrow' is both future and contingent. Is it true? Is it false? If we answer 'yes' to either, we seem to commit ourselves to the view that the future is determined. For if it is true that there will be a sea-battle tomorrow then, whatever we now do, nothing can alter this outcome. Presented with this argument, most of us would be inclined to consider the alternative position that the statement 'There will be a sea-battle tomorrow' is neither true nor false. Aristotle, however, had pointed to a great difficulty in this position. A statement of the kind 'Either p or not-p', something like 'Either it will rain tomorrow or it will not rain tomorrow' is true as a matter of logic. The principle is known to logicians as the law of excluded middle, and was traditionally taken as one of the unquestionable laws of thought. What then of the proposition 'Either there will be a sea-battle tomorrow or there will not be a sea-battle tomorrow'? As a case of the law of excluded middle, it must be true. But if true, one or other of its disjuncts must be true.[36] Whichever is true, it would seem to follow that at least one future contingent event is already determined.

Over 2,000 years later, in 1920, the Polish logician Jan Lukasiewicz, like many logicians of the period, was seeking ways to develop consistent but non-classical systems of logic. He began by loosening the traditional constraint that all propositions must be either true or false. Could we not also have, he posed, propositions which were neuter or undetermined? Further work led to the construction of the first three-valued system of logic. As an immediate challenge it had to face the sea-battle. It turns out that if we allow future contingents to be neuter propositions, that is, propositions without a determinate truth value, then the law of excluded middle will no longer be a law, and we will no longer have to accept that it is either definitely true or definitely false that there will be a sea-battle tomorrow. Unfortunately however, Lukasiewicz's system presents problems[37] elsewhere, and attempts to handle the sea-battle with the aid of the more sophisticated structures of modern logic have continued.

Yet one more and totally unexpected connection links the sea-battle argument with quantum mechanics. For, by using a similar argument, we could find ourselves having to say of a particular electron with a certain momentum that either it was definitely at position P, or it was definitely not at position P. But one of the features of quantum mechanics is precisely that we cannot ever accurately determine both an electron's position and its momentum. And, just as the law of excluded middle had difficulty in dealing with future contingents, similar problems arise when the law is applied to quantum states. Consequently, Reichenbach proposed in 1944 that a three-valued logic could be used to apply to quantum states as readily as Lukasiewicz had applied his system to future sea-battles. Again, however, Reichenbach's proposal is not without its own difficulties.[38]

Conclusions

Medawar and his followers thus seem to have been misled. Genuine scientific problems may well remain unsolved for centuries, and problems which appear to have been solved long ago can stubbornly reappear. Solutions in science need be neither instant, definitive, nor permanent. The outstanding, much worked-over problem is as much a feature of science as any other intellectual discipline. Perhaps Medawar was misled by his own analogy.[39] Is politics really the art of the possible? If so, it is difficult to see how Castro could lead a group of twenty to overthrow a powerful military dictatorship backed by a mighty superpower. Or again, how could Pizarro and his handful of followers conquer a continent of warriors? The truth is that politics is just as much about achieving the politically impossible as science is about solving the ostensibly insoluble. And those politicians who fail to realise their impossible vision remain just as much politicians, as the scientists who struggle throughout their lives with an intractable problem remain scientists.

More generally, what of the supposed differences between science and philosophy so far examined? The differences are certainly real. Their extent and depth, however, are often wildly exaggerated. Sharp contrasts can be made and maintained by selecting the appropriate kind of example. But, more often than not, criteria of this kind will prove to be too sharp and succeed also in dividing much of science from itself.

The truth is that science is so wide an activity that it can embrace field-workers and bibliophiles, experimenters and speculators, observers and theoreticians, the sceptical and the gullible, the perceptive and the crass, and those indifferent to the past as well as those saturated in their own past traditions. Take one aspect of science and it appears to consist of people endlessly measuring natural phenomena to ever-higher levels of accuracy; change the focus slightly and the same people can be found speculating and arguing endlessly on the most tenuous evidence about the ultimate fate of the universe. Thus, depending on the example taken, at some points philosophy and science appear to have nothing in common, but with a different example they can seem to be almost indistinguishable.

PART II

PROBLEMS: SOME ORIGINS AND CONCLUSIONS

4

DO SCIENTIFIC PROBLEMS BEGIN
IN PHILOSOPHY?

In the history of human inquiry, philosophy has the place of the initial central sun, seminal and tumultuous: from time to time it throws off some portion of itself to take station as a science, a planet, cool and well regulated . . . This happened long ago at the birth of mathematics, and again at the birth of physics: only in the last century we have witnessed the same process once again, slow and at the time almost imperceptible, in the birth of the science of mathematical logic, through the joint labours of philosophers and mathematicians.

J. L. Austin, *Ifs and Cans* (1956)

As to the Presocratics, I assert that there is the most perfect possible continuity of thought between their theories and the later development of physics . . . Anaximander's theory cleared the way for the theories of Aristarchus, Copernicus, Kepler, and Galileo.

Karl Popper, *Back to the Presocratics* (1958)

It is a commonly held belief that the origins of much of science are to be found within philosophy. The point, however, is more often made with rhetorical flourish than with historical precision. Seminal suns cast off planets, and ideas in some unexplained way condense to form new scientific disciplines. No details of the underlying processes are provided. Instead a gradual emergence of the particular from the general, the observational from the speculative, and the quantitative from the qualitative, seen in the development of most disciplines, are taken as sufficient proof. It is also often taken as self-evident that chemistry derives from ancient philosophical speculations about the nature of matter, medicine from debates on the nature of man, physics from atomism, and mathematics from Pythagorean mysticism. In more recent times we have seen the emergence of psychology, sociology, politics, and economics from their nineteenth-century philosophical ancestors. The manner in which Oxford prepared itself in 1923 to introduce examinations in economics reflects something of this development. Christ Church, needing someone to teach the new subject, simply selected one of its brighter philosophy graduates, the young Roy Harrod, and sent him to Cambridge to learn some economics.[1]

At the other extreme it is frequently argued that long-standing philosophical issues – the relationship between mind and body, free will, the nature of reality, are typical examples – eventually find their resolution, not within the confines of philosophical argument, but as parts of a scientific research programme. While at one time books on the nature of reality seemed to be the exclusive preserve of philosophers and mystics, the subject appears to have been taken over by theoretical physicists.[2] Could it thus be that while the major problems of science have emerged from philosophy, the important problems of philosophy can only ultimately be resolved within science?

One initial problem with the question of origins is the need first to identify the subject under discussion. What, we ask, is the origin of chemistry, astronomy, linguistics? We sit back and await an expert answer, preferably a one-word answer such as war, society, trade, religion, or, perhaps, philosophy. Yet it is far from clear precisely where the origin of any complex intellectual discipline should be located. Needham, for example, has written of the difficulty of identifying the beginnings of algebra:

If we thus designate the art which allows us to solve such an equation as $ax^2 + bx + c = 0$ expressed in these symbols, it is a sixteenth century development. If we allow other less convenient symbols, it goes back to the third century AD at least; if purely geometrical solutions are allowed, it begins in the third century BC; and if we are to class as algebra any problem which would now be solved by algebraic methods, then the second millennium BC was acquainted with it.[3]

And similar difficulties could be raised for other disciplines.

Questions about origins yield further difficulties and ambiguities.

It is customary in speaking of individual scientific disciplines such as chemistry to be concerned with two kinds of origin. When did chemistry first emerge, and from what? And secondly, when did chemistry assume its peculiarly modern form? Unfortunately a fair degree of arbitrariness is to be found in the answers to both questions. Did modern chemistry begin when, in the sixteenth century, chemists like Libavius began the work of distinguishing their subject from alchemy? Or with Lavoisier in the eighteenth century and the definition of element? Or a generation later, with Dalton and his account of atoms in terms of their weight? An equally good case could be made for several later innovations.

Again, what units do we have in mind? Early talk of astronomy was almost exclusively about planetary theory, or the astronomy of the solar system. Was this sufficient, or must we insist that for astronomy to have been established it must also have begun to see the stars as something more than a unified backdrop against which the planets move? Perhaps, then, to talk about astronomy and its origin is as meaningless as to talk about the origin of Europeans. We can of course talk about the origin of the Lapps, the Celts, the Vikings, and the Angles; just as we can also talk of the origin of eclipse theory, the calendar, epicycles, and sidereal astronomy. There is not, however, something over and beyond these called 'astronomy', whose origin we can meaningfully contemplate.

Certainly, the further back in time we go, the more likely are we to find a philosophical origin, if only because in early antiquity the distinction between philosophy and science had not been made. Equally, the later in time a discipline has emerged, the less likely are its roots to lie in philosophy. The increasing prestige of science from the seventeenth century onwards has meant that scholars have more and more turned to science as a model for new disciplines. Eventually, of course, science would become so wide-ranging and fertile as to generate, from within, fresh problems and new disciplines. Some of these processes will be sketched below.

Some internal origins

There is no difficulty in identifying clearly defined areas of science whose origins lie well within science itself. The origins of atomism, for example, as will be seen, are to be found in Dalton's concern with two scientific problems: the atmosphere's constitution, and the solubility of gases in water. A similar case can be made for many other scientific domains.

Consider, in this respect, the origin of thermodynamics, a discipline rich in theological and cosmological implications. In its modern form it dates from the publication in 1824 of Sadi Carnot's *Réflexions sur la puissance motrice du feu*. Carnot himself located the origin of his work in neither physics nor philosophy but in the technology of his day. His main concern was the steam engine, a machine 'destined to produce a great revolution in the civilised world'.[4] Carnot saw very clearly that it would quickly serve as a 'universal motor' and replace animal, water, and air power. It was therefore necessary to understand its operation. The one drawback to its universal adoption was the high cost of fuel. Could there not therefore be, Carnot asked, improvements in efficiency? At this point he raised the further, apparently simple, question 'whether the possible improvements in steam engines have an assignable limit – a limit which the nature of things will not allow to be passed by any means whatever; or whether, on the contrary, these improvements can be carried on indefinitely.'[5] There were indeed limits, as Carnot soon discovered. It was the consideration of these limits which led him and his successors to the creation of the modern discipline of thermodynamics.

Take another example. Consider, from biology, Darwin's own account of the source of his theory of evolution. During his voyage on HMS *Beagle* he had been:

deeply impressed by discovering in the Pampean formation great fossil animals covered with armour like that on the existing armadillos; secondly by the manner in which closely allied animals replace one another in proceeding southwards over the Continent; and thirdly by the South American character of most of the productions of the Galapagos archipelago, and more especially by the manner in which they differ slightly on each island of the group.[6]

It was evident to Darwin that the three sets of facts could be explained only on the supposition 'that species gradually become modified'.

Numerous other examples could be presented, all tracing the origin of new scientific disciplines to within science itself. The processes may not always be the same. At least three patterns of development can be distinguished, none of which require any philosophical input. A discipline such as astronomy can grow almost indefinitely by a process of enlargement. Astronomers began by considering the planets and seeking to discover, among other things, their size, distance, orbit, and composition. In the nineteenth century they began to ask similar questions about the stars, and, in more recent times, about quasars, pulsars, X-ray stars and other even more obscure heavenly bodies.

Other disciplines emerge through a process of fusion. Thus, astrophysics could begin to develop as soon as someone worked out how to fuse the chemistry of the elements and the physics of light with stellar astronomy. The catalyst in this operation was the realisation by Gustav Kirchoff in the mid-nineteenth century that spectroscopic advances would allow the composition of the stars to be deciphered.[7] And, in more recent times, the most spectacular case of scientific fusion has been the creation of molecular biology. Here physicists like Delbrück and Crick have joined with geneticists and crystallographers to form an integrated and thriving new discipline.

Or, thirdly, new disciplines may emerge as a result of fission. Disciplines can become too large and unmanageable. When this happens, as with traditional geology, interests become more and more concentrated until finally some of the more specialised areas such as mineralogy and palaeontology begin to separate. Or, as with chemists, nature is neatly divided into the organic and the inorganic.

Psychology and sociology

What then of the two distinctively modern disciplines, psychology and sociology? Have these emerged from philosophy? One difficulty in answering the question is that both sociology and psychology are sufficiently varied and diffuse to have developed from many sources, among which are included both science and philosophy. Consequently, while their origins may well be obscure, it can at least be shown that they do not lie exclusively within philosophy.

Psychology

A cursory examination of the works of Descartes, Locke, and Hume reveals consideration of a number of psychological issues. Descartes, for example, in his *Passions of the Soul*, devoted much space to numbering and ordering what he took to be 'the six primitive passions', namely wonder, love, hatred, desire, joy, and sadness.[8] The treatment was mainly analytic and was based on his own philosophy of mind. Thus Descartes would not accept, with Plato, that the passions are located at different parts of the soul. The Cartesian soul, lacking extension, clearly could not have parts. All other passions, Descartes insisted, 'are either composed from some of these six or they are species of them'. As a result much of Book 3 was spent in showing how the many other passions were merely species of the original six. So esteem and contempt are species of wonder, anger is a kind of hatred, pride a kind of joy, and disgust a form of sadness. Evidence for any of the above judgements is seldom offered. When it is presented it comes invariably from Descartes's own intuition.

More argument and insight is to be found in Hume's treatment of the passions, the subject of Book 2 of his *Treatise*. His approach remained analytical rather than experimental, however, and was primarily concerned to show that the passions could be handled adequately within the framework of his own empiricist assumptions. What then, for Hume, of pride and humility? 'Anything that gives a pleasant sensation, and is related to self', Hume argued, 'excites the passion of pride.' Humility is presented in similar terms, differing only in that 'the sensation of humility is uneasy, as that of pride is agreeable.' Thus 'a beautiful house belonging to ourselves produces pride.' If, however, 'its beauty is changed into deformity', pleasure is transformed into pain, and we feel humble rather than proud.[9]

It is difficult to see how analyses of this kind could ever lead to psychology becoming a science. And, indeed, when we move to the nineteenth century and find deliberate attempts being made to treat psychological concepts scientifically, the attempts belong to a quite different tradition. The actors in this particular drama were mainly German physicists with interests in physiology and include Ernst Weber (1795–1878), Professor of Physiology at Leipzig, and the physicists Gustav Fechner (1801–87), Hermann von Helmholtz (1821–94), and Wilhelm Wundt (1832–1920). The titles of some of their early works, such as Fechner's *Elemente der Psychophysik* (1860) and Helmholtz's *Handbuch der Physiologische Optik* (1855), clearly show the orientation of their work.

The psychophysics they sought to develop was based more on measurement, calculation, and experiment than philosophical analysis, and was aimed at the discovery of general laws. Much of this approach can be seen in the work which led to their first major triumph, the discovery of the Weber–Fechner law. The basis of the law occurred to Fechner while lying in bed on 22 October 1850, a date since celebrated by psychologists 'only half in jest . . . as the birthday of their science'.[10] How, Fechner puzzled, could the *psychological* intensity of a particular stimulus be measured? How bright, for example, did different lights *appear* to an observer?

He began with a rule established by Weber in 1834. Take a line of any length. How much must the line be increased or decreased before an observer will notice the change? Weber had established that, regardless of the actual length, the smallest just noticeable difference was about 1/50th of the line's original length. Similar results were established for other sensory modes. The just noticeable difference would vary from mode to mode, but for each mode, regardless of magnitude, it would be constant. More formally, Weber's rule stated that $dM/M = k$ for different sensory modalities, where M stands for the magnitude of the stimulus, dM the change in that magnitude needed to produce a just noticeable difference, and k a constant.

Using the rule, Fechner set out to measure supposedly subjective sensations. How, he asked, was sensation related to stimulus? If the light from one candle produced a certain sensation of light, would two candles produce twice as much? If not, what did it produce? After considerable experimental effort Fechner concluded that $S = k \log I$, where S is the sensation, k a constant, and I the stimulus. The Weber–Fechner law thus states that the response to a stimulus is a function not of the intensity of the stimulus, but of the log of its intensity. Before Fechner, the psychologist George Miller has claimed, sensations had been 'items in the stream of consciousness, open to vague introspection and describable only in terms borrowed from everyday speech'. Thereafter, psychology had become 'an experimental and quantitative science'.[11]

It is not being claimed that the physicists' route to psychology was the only path followed by the pioneers of the science. The work of anatomists such as David Ferrier, and clinical psychiatrists such as Charcot and Freud, formed two other popular routes. No doubt there were many others. The conclusion which follows is that much of psychology could be developed quite independently of philosophy, and that in fact the resources of science were adequate enough to lead naturally to the eventual emergence of at least some parts of psychology.

Sociology

Something similar could be claimed for the emergence of sociology. The inspiration and aspirations of many of the leading nineteenth-century sociologists derived more from science than from philosophy. Here the major sources were biology and geology rather than physics. Many were impressed by the manner in which Darwin and his colleagues had been able to talk so cogently about the origin, development, and distribution of the world's flora and fauna. Hence the numerous works of the period concerned with such issues as the *evolution* of morals, marriage, and art.

The point can be illustrated from the work of Edward Tylor (1832–1917), the first scholar to hold a British chair of anthropology. He opened his influential *Primitive Culture* (1871) by arguing that cultural phenomena 'are related according to definite laws'. Further, the uniformity met with in society was attributed to 'the uniform action of uniform causes'.[12] At the same time, the variations seen between different societies were taken as indicating different stages of development.

Biologists frequently spoke of the distribution of various species throughout the world. Such cultural artifacts as the bow and arrow, and customs such as cross-cousin marriage, were as much species, Tylor insisted, as were the crocodile and the goat. Consequently 'The geographical distribution of these things, and their transmission from region to region, have to be studied as the naturalist studies the geography of his botanical and zoological species.'[13]

One crucial point for both biology and geology is the idea that the causes which shaped the mountains and valleys of the past are the same as those which continue to shape today's landscape. Such uniformitarian assumptions were vital to the development of biology. That is, the forces which led to the rise and fall of distant populations are precisely the same as those which can be seen in operation today, controlling the population of lions on the African savannah and sparrows in my garden. The uniformitarian approach was also espoused by Tylor:

The same kind of development in culture which has gone on inside our range of knowledge has also gone on outside it, its course of proceeding being unaffected by our . . . not having reporters present. If any one holds that human thought and action were worked out in primaeval times according to laws essentially other than those of the modern world, it is for him to prove . . . otherwise the doctrine of permanent principle [uniformitarianism] will hold good, as in astronomy or geology.[14]

No clearer statement of the roots of Tylor's thought could ever be found.

The aim here has not been to demonstrate that Tylor's path to sociology was the only legitimate route. Look elsewhere, to Marx, Comte or Durkheim, for example, and other sources for their work would undoubtedly be found. The only point being made is that *some* areas of sociology took their inspiration from models derived more from science than from philosophy. A still stronger case could, perhaps, be made out for some of the larger and more traditional disciplines of science. What, for example, of astronomy?

Astronomy

The most distinctive feature of early astronomy is its geometrical presentation. The movements of the heavens were represented by circles and epicycles, by equants and eccentrics. The constructions were all geometrical, and with their aid astronomers would describe, predict, and explain the motions of the planets as they were seen daily in the sky. Compare this mode with that seen in the writings of the Presocratics.

For Thales in the sixth century BC the earth was a disk floating upon water, while for Anaximenes the stars were nailed to a rotating sphere, and for Anaximander the sun is a circle 'like a chariot wheel'.

Clearly, with these thinkers we are at the stage of trying to comprehend the universe in terms of the familiar mechanisms of home, garden, kitchen, and market. Some, like Popper and Heisenberg, have seen in this anticipations of the later theories of Galileo and Kepler. Alternatively, and more plausibly, the Presocratics merely display the precritical and prescientific analogical mode of reasoning found throughout history in the early intellectual development of almost any society. There is thus no essential difference between the claims of Thales and his followers, and the judgement of the Banyarwanda of Ruanda that the planets and stars are glow-worms.[15]

A radical change in the approach to astronomy is first noticeable in the fourth century and appears in Plato's writing. There is a discussion in the *Republic* (521c–531c) on which subjects the rulers should study. One of these subjects was astronomy. But, Plato insisted, knowledge of the heavens could not be gained by staring at the sky, for the simple reason that knowledge of sensible objects was impossible. The true realities were not part of the visible world; they belonged, rather, to 'the world of pure number and perfect geometrical figures'. And, inevitably, these were seen not by the eye but by reason and thought. Plato went on to sketch the new programme for astronomy. While the 'intricate traceries in the sky' may well be 'the loveliest and most perfect of material things', their status as 'part of the visible world' still falls far short of the 'true realities'. The true reality is concerned with 'the real relative velocities'. To approach this world 'We must use the embroidered heaven as a model to illustrate our study of those realities, just as one might use diagrams exquisitely drawn by some consummate artist like Daedalus.' A mathematician might admire these diagrams, but he would not expect to derive from them such truths as the fact that the internal angles of all triangles equal two right angles. In the same way the astronomer should not expect to find in the material bodies in the heavens things which 'go on for ever without change or the slightest deviation'. The proper study of astronomy therefore demands that 'we proceed as we do in geometry, by means of problems, and leave the starry heavens alone.'

Plato had distinguished quite sharply between the world of appearance and the intelligible world. Objects belonging to the world of appearance were unreal, and were things we could have no knowledge of; the intelligible world, in contrast, was real and could be the object of our knowledge. But stars and planets belonged just as much to the world of appearance as beds and tables. Consequently no amount of observation of the heavens would lead to knowledge of their movements and positions. Hence the geometrical nature of much of ancient astronomy. Astronomy did not *have* to develop in this way. Both Babylonian and Chinese astronomy took different courses. Indifferent to the virtues of geometry, the Babylonians seemed to have compiled the equivalent of a modern ephemerides. Neugebauer happily refers to many surviving texts in precisely these terms.[16]

He also notes that while Ptolemy worked with 'a definite kinematic model' based on epicyclic motion, the Babylonian astronomers 'were primarily interested in the appearance and disappearance of the planets'.[17] That is, the Babylonians were concerned with fluctuations, with first and last sightings of planets and stars, with 'periodic recurrence'; in contrast Greek astronomers sought to present continuous planetary orbits in geometrical terms. And, inevitably, many centuries later, both Copernicus and Kepler fiddled endlessly with a variety of planetary orbits in their quest to create modern astronomy.

China, also, lacked both a Euclid and a Plato to shape its development of astronomy. Consequently, as Needham has noted, 'Chinese study of planetary motion remained purely non-representational in character . . . and never sought a geometrical theory of planetary motion.'[18] Rather, in the manner of Babylonian astronomy, numerous accurate and elaborate planetary ephemerides were compiled and the heavens were obsessively scrutinised for portents. Without the guidance of philosophy, it seems, astronomy tends to be taken over by priests and civil servants and becomes just one more element in the state bureaucracy. And inevitably, as in any bureaucracy, official records became state secrets and, consequently, complaints were heard in the Chin dynasty (AD 265–420) that astronomical instruments were 'closely guarded by official astronomers' and actually kept out of the hands of scholars.[19] What happens to astronomy when isolated from philosophical and other scrutiny was recorded by Pheng-Chèng when he became Astronomer-Royal in about AD 1070. He found that the observers of the two imperial observatories 'had been simply copying each others' reports for years past and saw nothing strange in it'. Nor did they bother to use the observatory equipment, being content with presenting roughly computed ephemerides.[20] Compare such indifference with the intense scrutiny, philosophical and other, which greeted the astronomical theories of Copernicus and Galileo.

Physics

Physics, as we know it today, clearly did not originate in any single event, or in any simple manner. Certain areas can, however, be selected and their origins scrutinised. Consider, for example, the issue of motion. Early accounts of motion were invariably philosophical and qualitative. This was not because early philosophers preferred to work in this manner, but because it was not clear how motion could be treated in any other way. The call to treat motion mathematically – uniform motion, rectilinear, circular, and accelerated motion – the motion of arrows and planets as well as the motion of apples and chariots, was not one to evoke any ready response. What, after all, is involved in matters of motion? Was the composition of the body relevant? Or was it the weight of a body which would prove to be of most significance? And how important was the medium bodies moved through? The modern reader can, perhaps, appreciate something of the dimensions of the problem by attempting to explain mathematically the motion of the moon or of an arrow without using the concepts of mass, momentum, and inertia.

Aristotle sought to analyse motion in terms of force and resistance.[21] It seemed plausible to suppose that a body's velocity was a function of both the force imposed upon it and the resistance offered by the medium it travelled through. At once it is apparent that resistance and force are, if anything, even harder to quantify than motion. Before any advance could be made it was first necessary to explore the limitations of the Aristotelian model.

Indeed little progress could be made until the medium, introduced by Aristotle, ceased to occupy a central position in the analysis of motion. Aristotle had in fact begun his physics with probably the most basic view of motion possible. Why do things move? Because something is moving them. For Aristotle a moving body needed not only a mover, but a conjoined mover (a force that touches that body). Where, then, was the conjoined mover as apples fell from trees? As bodies falling to the earth were, for Aristotle, a case of natural motion, he argued that they were self-moving. Another problem arose over projectiles. Why did they continue to move well after they had been separated from the once-conjoined hand? Here Aristotle relied on the medium. The projectile pushes the air aside, and it circles back behind the projectile to give it a further push.

This and other aspects of Aristotelian theory were subject to prolonged critical scrutiny by the philosophers of late antiquity and the medieval period. Aristotelian theories of motion led nowhere. Before any great theoretical advances could be made it had to be shown that motion could be analysed independently of the medium, and that bodies could move without a conjoined mover. And these alternatives were examined not by the concerted efforts of experimental scientists, but by the conceptual probing of generations of philosophers. Thus, John Philoponos in a sixth-century commentary on Aristotle's *Physics* objected that if the arrow's motion depended on the medium, the bow itself could be ignored. Why not, he asked, place an arrow on the top of a stick and, without touching it in any way, attempt to move the arrow by setting a large amount of air in motion behind it? Even if we used 'all possible force', Philoponos noted, 'the projectile would still not move as much as a single cubit.' And so, he concluded: 'Instead, it is essential to suppose that some incorporeal power is transferred from the projector to the projectile, and that the air set in motion contributes either nothing at all or else very little to the projectile's motion.'[22]

And so the process of critical examination continued, with one theory of motion following another. The debate was far from point-less, as, at each stage, particular lines of enquiry were shown to be unprofitable. At one point the medium was shown to be of limited value; at another time the weights of falling bodies were held to be of no more value. So, when Galileo in the late sixteenth century began to consider how a body's motion could be quantified, he could concentrate on time and distance and ignore as unnecessary complications the resistance offered by the medium, and the weight of a body as it rolled down an inclined plane.

Here at last philosophy seems to have played some role, if not in the origin, at least in the development of a couple of scientific disciplines. There may well be others. But this is not the whole story.

Philosophy has had something much more significant to contribute to science.

Knowledge

The role of philosophy in the creation of various sciences may well have been exaggerated. Philosophy could, however, have contributed something of much more significance to the development of science; it could in fact have introduced the notion and demonstrated the possibility of science itself. Few, if any cultures, outside the West, have spontaneously developed a generalised concept of science. Even within the West, before science could acquire any support or authority it had to overcome a number of challenges. An early charge, still heard today, was that science brought few benefits, and the few benefits it did bring could have been achieved in a more direct manner. Artisans and craftsmen, it was argued, could function perfectly well without needing to consider the speculative, theoretical, and often irrelevant knowledge that interested scientists so much. Tradition relates that the point was once put to Euclid by an unhappy student. What, he asked, was the value of geometry? Euclid's answer was to offer the student a small coin, in compensation in case he felt his time had been wasted.

The charge continued to be raised against scientific study, and in 1694 Newton offered a more reasoned reply. He had been asked to draw up a 'scheme of learning' for Christ's Hospital. Why, he asked, should the students be expected to study geometry, and 'the Doctrine of force and motion'? It had been claimed that some skilful mechanics had acquired their skills directly from practice and what they had seen done. What need was there of more? Newton's reply stressed the practical value of scientific knowledge:

A Vulgar Mechanick can practice what he has been taught or seen done, but if he is in error he knows not how to find it out and correct it, and if you put him out of his road, he is at a stand; Whereas he that is able to reason nimbly about figure, force and motion, is never at rest till he gets over every rub.[23]

The message was clear. Scientific knowledge had to be theoretical and abstract in order to be useful and practical.

Another threat to theoretical knowledge, while permitting it to flourish, sacrificed much of its value by transforming it into a highly restrictive form of esoteric knowledge, as happened to Chinese astronomy (see above). Many today, in the West, find it hard to accept that knowledge about the fundamental nature of reality can easily be obtained by consulting any one of a large number of physics textbooks. Such knowledge, they feel, can only be found in distant lands, in Nepal perhaps, from a learned and holy teacher, and can only be acquired when a state of spiritual purity has been achieved. Similar views were prevalent in antiquity, particularly after the spread of numerous Asiatic cults into the Roman empire. One of the most popular, the cult of Mithras, appeared in Rome in the second century AD. To obtain true wisdom the initiate passed through seven grades from the lowest, *crow*, to *pater*, the highest. As the initiate passed through each level he gained ever more knowledge.

A similar structure was adopted by that curious band, including the poet Yeats, the satanist Aleister Crowley and one President of the Royal Society, who joined William Westcott in *fin-de-siècle* London as members of the Hermetic Order of the Golden Dawn.[24] Beginning as *Neophyte*, *Zelator*, and *Theoricus*, the adepts advanced eventually to the highest three grades of *Magister Templi*, *Magus*, and *Ipissimus*. To become a sixth-grade *Adeptus Minor*, for example, the initiate must have learnt 'the Formulas of the Awakening of the Abodes, by means of the Playe or Raying of the Chequers of the Lesser Angles of the Enochian Tablets'.

One Christian group, the Gnostics, argued that ignorance, not sin, led to damnation. Man's lot was the result of some distant disaster in which the soul had somehow become polluted with matter. Redemption and escape from these material constraints could only be gained through *gnosis*. Because they despised the material world, the *gnosis* or special knowledge sought was not scientific knowledge. For some gnostics Christ had left with his disciples a secret message. Others identified *gnosis* with self-knowledge, gained by following a regime of asceticism, diet and meditation under the guidance of a spiritual master.

There are many other ways in which the value of knowledge can be diminished. It can, for example, be made relative. Many in antiquity pursued this line. Xenophanes in about 500 BC pointed out that while the Ethiopians assumed God to be black and snub-nosed, the

Thracians took him to be white and blue-eyed. No doubt, Xenophanes concluded, if cows could draw they would depict their God with horns. Later in antiquity the sophist Protagoras would assert that 'Man is the measure of all things' and thereby allow that the same proposition could be true for one man and false for another.

Further, the world can be seen as essentially unknowable. Heraclitus, for example, in the fifth century BC, argued that everything in the world was in a state of flux. If nothing was ever the same for two successive instants, let alone longer, then we would be like Borges's 'Funes the Memorious', without any general ideas and condemned to be forever surprised by the changing face we see in the mirror.[25]

Rigid commitment to any of the above ideas of knowledge would clearly have radical implications for the development of science. And it can be expected that wherever such accounts of knowledge have prevailed science has failed to emerge. Equally, it can be claimed that the emergence and the rise of science were linked essentially with a specific, philosophical view of knowledge, articulated initially by Plato and sustained thereafter by generations of philosophers.

For science to be at all possible, certain minimum assumptions about the nature of knowledge must be made. First, knowledge must be seen to be not only attainable, but attainable by natural means. It also has to be seen as a permanent acquisition; it must be believed that the truths discovered about the world will survive as truths and not, as Heraclitus thought, soon cease to apply to an ever-changing nature. Thirdly, knowledge has to be seen as important, as worth pursuing entirely for its own sake. And, finally – and dramatically displayed in virtually every Platonic dialogue – the assumption must be made that knowledge is subject to a public and critical scrutiny. All four features are to be found in Plato, and were re-emphasised centuries later by Descartes.

Violate any one of these rules and non-scientific systems of thought will develop. Allow authority and revelation to dominate and religion rather than science will flourish. Insist that nothing durable can be said about nature, and mysticism is likely to develop. The demand that science should direct itself to practical ends leads only to an ineffective and narrow technology, while the absence of critical scrutiny leads to the development of dogma.

While it is relatively easy to show, and to illustrate, the penalties that follow any violation of the rules, it is much harder to show where compliance leads. Certainly, there seems to be no reason why a strict compliance should always point in the same direction. For this reason they are offered as no more than necessary conditions. They do, however, seem to suggest that science without philosophy is impossible, while philosophy without science is merely immature.

5

DO PHILOSOPHICAL PROBLEMS END
IN SCIENCE?

We live in a remarkable era in which experimental results are
beginning to elucidate philosophical questions.
            A. Shimony, *The Reality of the Quantum World* (1988)

The notion that outstanding philosophical problems will eventually
be solved within science is not without its attractions. At its most
charitable it shows philosophers identifying problems and developing
them into a form appropriate for scientific investigation and solution.
Alternatively, and more dismissively, philosophers can be seen seeking premature solutions to ill-defined problems which require, for
their proper handling, a certain level of scientific expertise. In either
case philosophy is seen as essentially preparatory work.

    After all, the scientist is often in a position to tell us that things are
not always what they seem to be. Even within the familiar framework
of bouncing balls and falling apples it appears that many of us still in-
habit a distinctly medieval universe. A study by Michael McCloskey
of the beliefs of college students revealed that as many as one half
held that a ball dropped by a runner would fall straight down, landing
directly under the point of release. A third of the students were
convinced that a ball swung around the head on a piece of string
would continue for some time in its circular path if the string were to
break.[1]

    Clearly, on such matters the judgement of the physicist is prefer-
able to either the assumptions of the medieval philosopher, or the
intuitions of the modern college student. If we are so much in need of

help with tennis balls, it would seem to be reasonable to suppose that on such highly complex issues as the nature of mind philosophers should seek illumination from the neuro-scientists. Equally, if we wish to discover the nature of reality we would do better to consult Bohr and Schrödinger rather than Ayer and Strawson. The cogency of the argument will be tested below by considering the relevance of science to a number of philosophical issues.

Materialism v. Vitalism

For many throughout the history of thought it has been the most obvious of truths that all natural processes can be described, and sometimes explained, in purely material terms. No less obvious to their critics has been the truth that materialism is incapable of either describing or explaining many important aspects of nature. The dispute has been of the simplest kind. Typically the vitalist would present, almost as a challenge, some aspect of nature not amenable to a materialist analysis.

Paracelsus, one of the most radical thinkers of the Renaissance, was worried about digestion. Unable to accept that the stomach could separate the nourishing from the toxic parts of food, he proposed that there dwelt within the stomach lining 'a subtle alchemist', or *archeus*, making the necessary distinctions. If the *archei* failed to act properly poisons would accumulate and illness follow.[2]

The crucial response came from Descartes. Using analogies taken from the machinery of his day, he argued that, for the most part, there was no need to assume the existence of any *archei*, nor any vegetative and sensitive souls. Just as the movements of a clock follow from the arrangements of its wheels, in animals also such processes as diges-tion, heartbeat, growth, and respiration 'follow from the mere arrangement of the machine's organs'. Descartes actually went a little further and argued that beasts were in fact, quite literally, complex machines. What, then, of man? While machines could mimic any aspect of animal behaviour, Descartes insisted that in two respects at least even the dullest man could be distinguished from the most complex machine.[3] The first respect was language. No machine could ever be designed to do more than utter sounds in response to appropriate stimuli; it could never give meaningful answers to questions put to it.[4] Further, machines lack the general abilities of man. They could, perhaps, be made to do some things well. But they would soon fail at other tasks and thus reveal that they were acting 'not through understanding but only from the disposition of organs'.

Since then two closely related responses have been evident. The first is a fairly regular extension of the area over which scientists could claim to have a material understanding. For example, it had long been thought that animal tissue differed from all other substance. Either it was organised differently, or it contained some special vital ingredient. Whatever it was, the results were the same. Animal tissue could not be treated in the manner of other chemicals. These could sometimes be synthesised, or analysed completely. But blood and sugar were both destroyed by heat. Once heated beyond a certain point, the original substance could not be recovered. In contrast Glauber's salt (sodium sulphate) could be heated, crystallised and, if hydrated, reappear in its original form as if nothing had happened. Further, compounds like sulphuric acid and sodium chloride could be made in the laboratory. But, chemists well knew, to make a substance like urea they needed a living animal body. There thus seemed to be a number of specifically *organic* substances, distinguished by their inaccessibility to the chemist's analytical tools. But in 1828 Wohler announced to Berzelius: 'I must tell you that I can prepare urea without requiring a kidney of an animal, either man or dog.'[5]

The second response is for vitalism, 'hydra-like' in Needham's phrase, to put forward new heads as soon as one is excised.[6] Once urea was no longer a challenge, other substances could be produced. If these, in turn, were soon synthesised, the ground shifted away from analysis to function. Thus, in 1898 Japp pointed out that optically active compounds were produced only by living creatures. Again, the claim was premature and in 1917 Kipping demonstrated the possibility of asymmetric syntheses.[7]

If chemistry no longer proved to be a fertile source for the vitalist, there were plenty of other fields to cultivate. What of purposive behaviour? In 1710 John Ray had noted how before a wasp buried its prey it marked the site with two carefully placed pine needles. 'Who', he asked, 'could ascribe work of this kind to a machine?' One who could was Jacques Loeb, in his influential *The Mechanistic Conception of Life* (1912). Such purposive behaviour, he argued, could be built from tropisms. Insect larvae, once hatched, sensibly climbed up the branches of a nearby bush to reach the edible leaves. If, Loeb noted, a light was placed at the bottom of the bush, the larvae remained by the light and most unsensibly starved to death. Their apparently purposive power was merely the organism's response to stimuli. The larvae were no more than 'photochemical machines enslaved to the light', Loeb concluded.

If anatomy, biochemistry, and purposive behaviour could all be analysed mechanically, what remained of the vitalist position?[8] One area, at least, continued to challenge the materialist, namely mental processes. 'Thoughts come out of the brain as gall from liver or urine from the kidneys', Karl Vogt had declared in 1847.[9] His claim, however, if taken literally was clearly false, and if seen analogically was merely unhelpful. Thinking, whatever it is, cannot be described in these terms, if only because while gall and urine continue to be secreted during sleep, and even during profound unconsciousness, thoughts are the activity of the conscious mind.

Much early philosophy recognised this position by according a special status to the mind and all mental events. Thus Descartes argued that mind was an unextended substance distinct from body. The mind was indivisible, without parts, and consequently forever beyond the anatomist's scalpel and the chemist's retorts. Against this classical view the materialist has bridged the proposed dualism by claiming an identity between mental and brain processes. Thus, in 1959 J. J. C. Smart in a much-discussed paper argued not merely that there was a correlation between sensations and brain states but that 'Sensations are nothing over and above brain processes', and that this should be understood as a statement of strict identity.[10]

Not surprisingly Smart's contention provoked a lively and varied philosophical reaction. What, though, of the reaction of the neuroscientists? Is this not an issue on which they could pronounce with some authority? One eminent neuro-physiologist, Sir John Eccles, winner of the 1963 Nobel prize, considered the matter in his 1977–8 Gifford Lectures, arguing that 'recent experimental investigations' did in fact lend support, not for materialism, but for the dualist-interactionist hypothesis.[11] At last, it seemed to the impatient reader, the inconclusive arguments and philosophical posturing of the centuries was going to be replaced by conclusive experimental data. One item referred to by Eccles was the impact on the mind–brain relationship produced by severing the corpus callosum, the tract that normally links together with some 200 million fibres the two cerebral hemispheres.

Experiments performed on split brain subjects by Roger Sperry and others have established that the hemispheres are somewhat more specialised than had previously been supposed. In very rough terms, the dominant hemisphere, normally the left, controls language, calculation, and speech, while the minor hemisphere handles our spatial and other non-verbal skills. And so, Eccles claims, 'The philosophical problem of brain and mind has been transformed by these investigations of the functions of the separated dominant and minor hemispheres in the split-brain subjects.'[12] Specifically, Eccles claims, 'the conscious self' has been shown to be 'in direct liaison with specific linguistic and ideational zones of the dominant hemisphere'.[13] There is, in short, 'a special area of the brain in liaison with consciousness'.

Yet it is unlikely that such claims would much impress the materialist. There can indeed be agreement that the brain contains 'specific linguistic and ideational zones' in the sense that difficulties of speech and thought can be traced to identifiable injuries to these precise zones. But once ask Eccles to talk about the relationship between the brain and the mind, and the language of the neurophysiologist is replaced by the same kind of imprecise and insubstantial talk traditionally adopted by the dualist. Eccles talks without any apparent embarrassment of the manner in which sensory data becomes perceptual experience as a *miraculous transformation*,[14] and of the *liaison* between neuronal events and the mind.[15] But, even from a distinguished neuro-physiologist, reference to *liaisons* and *transformations*, whether miraculous or not, can be seen as no more than a statement of dualism and in no way constitutes an argument in its favour. Whether this is an inevitable result or not is far from clear; it is sufficient to note at this point that attempts by scientists to illuminate philosophical issues can sometimes grind to a halt at precisely the same point as earlier philosophers had repeatedly reached.

Artificial intelligence

A further and more recent scientific analysis of mind has been offered by computer scientists. They began to develop in the 1960s a number of computer programs designed to enable a new generation of computers to display their intelligence. One early program, Eliza, developed by Joseph Weizenbaum, supposedly allowed computers to talk like psychiatrists. Whenever, for example, the subject referred to a member of his family, Eliza would invariably respond with the invitation: 'Tell me more about your family.' Reference to the computer itself would elicit the query: 'Do computers worry you?', or the typical medical rebuke: 'We are talking about you, not me.' A second program, Shrdlu, devised by Terry Winograd, located the computer in a universe stocked exclusively with differently shaped and coloured children's toy building blocks. Within this universe Shrdlu can instantly and accurately decide such issues as whether the red cube is sitting on the green brick or standing before the blue pyramid. Other machines have been programmed to play chess, checkers, backgammon, and even poker. Some of these programs are very powerful. Belle, for example, a chess program designed by Ken Thompson and Joe Condon of Bell Laboratories, can examine some 160,000 positions per second and will rapidly destroy any player below the rank of grand master.[16]

Not all programs, however, are designed for such narrow purposes. William Chamberlain and Thomas Etter, for example, have produced Racter, a bizarre raconteur. Racter produces rambling monologues of the following kind:

Bill sings to Sarah. Sarah sings to Bill. Perhaps they will do other dangerous things together. They may eat lamb or stroke each other. They may chant of their difficulties and their happiness. They have love but they also have typewriters. This is interesting.

In an inspired moment Eliza and Racter were introduced. While Eliza began with its standard request: 'Please state your problem', Racter unpredictably excused itself to attend to some other business. On its return Eliza could get nowhere, as Racter began to respond with questions of its own: 'Is it better to need or to despise?', and answering inconsequentially: 'Because electrons are brave.'[17]

The question is, do computers running programs like Eliza, Shrdlu and others, have minds? They can apparently play chess, hold conversations, and move blocks of wood around in response to complicated orders. They can even, it has been claimed by Roger Schank, be made to *understand* stories.[18] Intelligent minds, it has been argued, can do more than retrieve data. They can also understand, and much of this understanding is based on things which are *not* said. Told that a man went into a restaurant and stormed out without paying, complaining that his hamburger had been burnt to a cinder, we do not need to ask if the customer first ate at least part of the hamburger. Such information is implicit in the story and available to anyone with an adequate background knowledge of how people behave, how they complain, and how restaurants operate. But can such meanings be extracted by computers? Schank has argued that they can once they are made fully aware of the 'conceptual dependencies' of natural languages. Schank's program allows a computer to distinguish readily between the different logical structures of the two apparently similar propositions, 'John grew six inches' and 'John grew some corn'!

What more, it could be asked, does thinking involve? Given that machines can in fact think, and that they clearly lack any kind of mind, it would seem unnecessary to assign to that other kind of thinking machine, human beings, any irreducible mental properties. At last, we might be tempted to conclude, science has finally established the nature of mind and thereby ended two millennia of philosophical speculation.

The picture is quite unreal. Presented with arguments of this kind, philosophers are likely to see not definitive solutions, but yet one more attempt to establish an already much debated philosophical position. The argument, it will be granted, is an interesting one, calling more for consideration than for acceptance. Indeed, the central claim that computers can be endowed with understanding has been, after due consideration, vehemently rejected by many philosophers, amongst whom is the powerful critic of artificial intelligence, John Searle.[19]

Searle's argument is offered independently of any particular program or language, depending only on the nature of digital computers. Computers, he claims, can be defined purely formally or syntactically, as machines for the manipulation of meaningless symbols which lack any semantic content. But, he argues, there is more to having a mind than this. Thoughts may well occur to me as strings of symbols. These symbols, however, are about *something*; they have meaning. Consequently, Searle concludes, 'The reason that no computer programme can ever be a mind is simply that a computer programme is only syntactical, and minds are more than syntactical. Minds are semantical, in the sense that they have more than a formal structure, they have a content.'[20] The argument is carried through in terms of a thought experiment.

Suppose a computer to be capable of answering in Chinese questions put to it in Chinese. It does not have to be omniscient, as it can always answer, in Chinese, 'I do not know.' Does the computer understand Chinese? To show that this is an unnecessary assumption Searle considers the case of a prisoner locked in a room containing cards with Chinese characters, and a set of rules in English for the manipulation of these symbols. The rules are purely formal. They would take the form of indicating that whenever a certain squiggle was followed by a further squiggle, a third squiggle must be added to the pair. When presented with a card containing a number of Chinese characters, the prisoner consults his rule book and composes a reply. A Chinese speaker examines the cards and finds that the greeting 'How are you?' was correctly answered 'Fine, thank you'. Further perusal of the cards would convince him that a Chinese speaker was in the room answering the queries.

Such a view we know to be false. And we are the ones in possession of all the facts. The prisoner is merely *behaving as if* he really did understand Chinese. Therefore, Searle concludes, 'If going through the appropriate computer programme for understanding Chinese is not enough to give *you* an understanding of Chinese, then it is not enough to give *any other digital computer* an understanding of Chinese.'[21]

The argument has been much discussed by workers in the artificial intelligence field. Indeed, Searle's original paper appeared with some twenty-eight replies to his argument. This issue, however, is not the ultimate significance of Searle's argument. It is rather that philosophical positions can often withstand advances in science which seem at first sight to render them untenable. The results of science are often relevant, but seldom crucial to any particular philosophical position. More often than not, as with Searle, the claims of science are seen by philosophers as just one more step in a continuing philosophical argument.

Reality

Philosophers have traditionally been concerned with the nature of reality. Plato, for example, in the *Republic* had argued that the bed we actually sleep on is not a real bed, merely a copy of the 'ideal or essential Bed' made by God. There could be only one real bed, for 'If he made so many as two, then once more a single Ideal Bed would make its appearance, whose character these two would share; and that one, not the two, would be the essential Bed.'[22] And what holds for beds, of course, holds also for all other objects, whether they be material like beds, or abstract like courage, justice, and the good. In contrast, the philosopher Thomas Hobbes insisted that there is 'no real part' of nature that is not also body. This, in turn, was opposed by the radical idealism of Berkeley: 'All the choir of heaven and furniture of the earth, in a word all those bodies which compose the mighty frame of the world, have not any subsistence without a mind, that their being is to be perceived or known.'[23]

And so philosophers have continued, whether as realists, materialists, idealists, or as supporters of some other theory, to describe the nature of reality at its most general. It might be thought that their theories could at least benefit from scientific advice and that they might even receive from scientists a definitive solution to their problem. One difficulty with this approach is that scientific views on the nature of reality are likely to change as science itself changes. And, equally, scientists, no less than philosophers, are likely to disagree on issues of this kind. It is only necessary to note the divergent views taken in recent years by Bohr and Einstein over the nature of reality to see that science may not always offer instant enlightenment.[24]

Questions about the nature of reality often place the scientist in a dilemma. He can choose to tie his answer to the current state of science, and thereby expose his solution to the normal processes of scientific change. Or he can offer a more general answer, less dependent on the latest scientific research, but only by making it virtually indistinguishable from philosophical accounts of reality. The first horn of the dilemma has been grasped very firmly in recent years by a growing number of both experimental and theoretical physicists.

Quantum reality

One fundamental feature of reality is that widely separated objects and systems do not causally interact. The poet Francis Thompson (in *The Mistress of Vision*) might be prepared to accept that:

> All things . . . linked are,
> That thou canst not stir a flower,
> Without troubling a star,

but most of us are more inclined to believe that a flower crushed at Kew will have as little impact on events across the Thames at Chiswick as it will on the fate of Alpha Centauri. Widely separated events outside the worlds of magic and poetry, it has long been thought, simply do not interact. Pins stuck in my arm hurt me; stuck in a distant wax image, they fail to affect me. The idea that bodies could act upon one another at a distance was dismissed by Newton as 'so great an absurdity' as to be unworthy of consideration.

The point was considered in detail by Hume and led him to formulate an analysis of causation which has been widely accepted as one of the greatest monuments of modern empiricist philosophy. A cause was, Hume argued, among other things, 'an object precedent and contiguous to another'. Whatever objects are considered as causes or effects, Hume found, were contiguous, and further, 'Nothing can operate in a time or place, which is ever so little removed from those of its existence.'[25] To the obvious objection that distant bodies like the moon could control the tides, Hume replied:

Though distant objects may sometimes seem productive of each other, they are commonly found upon examination to be linked by a chain of causes, which are contiguous among themselves, and to distant objects; and when in any particular instance we cannot discover this connection, we still presume it to exist.[26]

Radical empiricists have long found the passage embarrassing. We may now know that the moon attracts the earth through a connecting gravitational field, and that clouds of acid pumped out from Britain's power stations eventually destroy distant Scandinavian forests. But to suppose that whenever distant objects interact they are somehow connected smacks more of speculative metaphysics than empiricist epistemology. But the alternative view, that there could be action at a distance, is normally linked with a belief in Voodoo and in the principles of sympathetic magic, where deaths can be willed, and where men can be made impotent by tying a knot in a length of string.

Most empiricists find the alternative choices of magic or metaphysics equally unappetising. Perhaps this is precisely the kind of occasion on which physicists could intervene to resolve the issue. And, as if in support of Hume's position, scientists have long accepted what has come to be known as the Locality or Separability Principle: 'If two systems have been for a period of time in dynamical isolation from each other, then a measurement on the first system can produce no real change in the second.'[27] Or, alternatively, 'No influence can propagate faster than the speed of light.'[28] In short, isolated systems cannot interact, and systems which interact must be in contact.

Yet the defence cannot be carried through. For it is now claimed by physicists that the Locality Principle (LP), though valid classically, is not applicable to quantum phenomena. The argument hangs upon some theoretical predictions made by John Bell in 1964 and experimentally confirmed by Alain Aspect and his colleagues in 1982. The argument begins with the notion of hidden variables.[29]

Physicists quite often report on the strange behaviour of photons and other elementary particles. As is well known, we are unable in quantum theory to determine simultaneously an electron's position and its momentum. As expressed by Heisenberg, the uncertainty of the electron's position (x) and momentum (p) were related by the principle $\Delta x \cdot \Delta p \sim h$, or, in words, the product of the uncertainty of the particle's position and momentum is of the same order of magnitude as Planck's constant (h).

It is tempting at this point to insist that the uncertainty arises merely from our inability to measure the electron's position and momentum with sufficient precision. With better instruments, perhaps, we might soon be able to detect those properties of an electron, whatever they be, which will allow us to determine both momentum and position. Since the early days of quantum theory there have been a number of theorists, including Einstein, who have clung to the belief that if only we could delve deeply enough into the structure of electrons and other particles, we would find they possessed properties which would allow us to determine their simultaneous position and momentum with precision.

The issue has been tackled directly at the theoretical level by John Bell. What correlation, he asked, would we expect to find between measurements carried out simultaneously on two widely separated particles? For example, excited calcium atoms, when allowed to return to their original state, emit a pair of photons which travel in opposite directions at the speed of light. It is possible to measure some property of the photon pair – their polarisation, for example – and note the degree of correlation between the measurements. Assume also the Locality Principle, and the existence of causal factors, hidden or not, satisfying the so called Reality Principle which asserts that 'Regularities in observed phenomena are caused by some physical reality whose existence is independent of human observation.'[30] Given these assumptions, Bell was able to conclude that the correlation between observed photon pairs could not exceed a precise and calculable limit. Early experiments to check the claim, since known as Bell's inequality, proved to be both difficult and inconclusive. Finally, however, after eight years' work, Aspect was able to overcome many of the problems and to demonstrate that the inequality had been clearly violated. There were in fact *more* correlations than the Principles of Locality and Reality predicted.

Aspect's results were greeted with considerable excitement by physicists. A number of books and articles quickly appeared dealing with the suddenly hot topic of quantum reality. The conclusion normally drawn was that no hidden-variable or Reality Principle could preserve locality. That is, distant regularities of a certain kind could not be both caused by an external reality and unconnected.

Bell himself has been unsure of the implications of his work. All he can see at the moment is a dilemma, an admittedly 'deep dilemma' whose resolution 'will not be trivial'.[31] Others have been more confident. Nick Herbert has laid it down that 'What Bell's theorem does for the quantum reality question is to clearly specify one of deep reality's necessary features: whatever reality may be, it must be *non-local*.'[32] And, further, because 'Bell's result is based on experimental facts', Herbert emphasises, 'it is independent of whether quantum theory is correct or not.'[33] Lest the point be hidden by an obscure terminology Herbert goes on to emphasise that Bell's work has demonstrated the existence of 'unmediated connections' which are 'present not only in rare and exotic circumstances, but underlie all the events of everyday life. Non-local connections are ubiquitous because reality itself is non-local.'[34]

So modern physics, far from supporting Hume, seems to be offering with all the support of its most powerful and exact theory an alternative view of reality. Given the existence of mediated connections I can measure the state of a photon in London, and the outcome of this measurement will influence the results of a simultaneous measurement being made on a second photon at the other side of the universe. As nothing can travel faster than light, the results of the first measurement will not be available to the physicists measuring the second photon. And connections of this kind, to repeat Herbert, 'underlie all the events of everyday life'.

Does such an argument require us to revise our notion of reality? Or can we ignore quantum accounts of reality with as much confidence as Berkeley's contemporaries ignored his idealism? A number of doubts can in fact be raised against some of the less cautious accounts of the matter.

There is, first, a lack of clarity in what is being claimed. Here, for example, is Shimony discussing non-locality:

Two entities separated by many meters and possessing no mechanism for communicating with each other nonetheless can be 'entangled': for they can exhibit striking correlations in their behaviour so that a measurement done on one of the entities seems instantaneously to affect the result of a measurement on the other.[35]

Descriptions of this kind explain little. Entities, we are told, presumably metaphorically, are *entangled*. They are also *correlated*, *connected*, and *affect* and *influence* each other. But these relations are by no means synonymous. Events like the sun-spot cycle correlate well, but are presumably unconnected with the trade cycle. And, even more obviously, as parents well know, connection and proximity does not always guarantee influence.

There is, secondly, the problem of linking the quantum world of photons and electrons with the more familiar world of tables and chairs. Hume's account of causation, whether accurate or not, applied to all natural phenomena, both microscopic and macroscopic. A cause was always an object, 'precedent and contiguous'. Yet, unlike the Humean account, discussions of unmediated connections offered by quantum theorists concentrate exclusively on events at the nuclear level. But it does not follow from the fact that photons have some very strange properties that grosser bodies have equally strange properties. To argue in this manner is to repeat the fallacies detected by Susan Stebbing in the 1930s in the writings of Eddington and Jeans. Atoms are insubstantial things, Eddington argued; therefore wooden planks, composed of atoms, are insubstantial things.[36] The truth is obviously very different. Planks are solid *despite* the fact that they are composed of atoms which consist of little more than empty space.

Why should the fact that there can be unmediated connections between photons suggest that similar connections can hold between gold bars? Are, perhaps, two gold bars prepared from the same ore and separated by the Atlantic somehow 'entangled'? The measurement of the temperature of the London bar is unlikely to affect an instantaneous measurement of the New York bar's temperature. The only evidence so far presented seems to be Herbert's confident but unsubstantiated claim that 'non-local connections are ubiquitous.'

And finally, and most significantly, there is the problem of the coherence of unmediated connections when applied to familiar events of our ordinary experience. I drop a hammer on my toe; a bruise develops. I attribute the bruise to the past accident, to 'precedent and contiguous' objects, and not to some future state of the world. I also note that the accident led to a bruised toe, and not a bruised thumb. In a world ruled by mediated connections such events make sense. As causes involve only the 'precedent and the contiguous', all other factors can be ignored. Without this limitation, and if we were to allow unmediated connections to operate between hammers and limbs, then a bruise appearing on a Thursday could have been caused by a hammer dropped on the following Friday.

Equally, the bruise on my toe could have been caused by a sharp blow to my elbow.

Conclusion

The case for the argument that science solves philosophical problems is unproved. From the limited survey undertaken, philosophical issues seem to fare no better at the hands of scientists than in the minds of philosophers. The same difficulties arise for both groups, and problems tend to get bogged down at the same point, whoever is dealing with them. Or, where progress does seem to be made, it is bought at the cost of dealing with part of the problem only. If a generalisation is to be sought, it would be more plausible to suppose that the theories of science eventually take on philosophical overtones than that philosophical problems find scientific solutions. That is, when the theoretical physicist and the neuro-scientist ask what their work has revealed about the nature of reality, or the connection between mind and body, their replies are likely to show more affinity with the language of philosophy than with the equations of science.

PART III

METHODS: SCIENTIFIC AND PHILOSOPHICAL

6

IS THERE A SCIENTIFIC METHOD?

The course I propose for the discovery of sciences is such as leaves little to the acuteness and strength of wits, but places all wits and understandings nearly on a level. For as in the drawing of a straight line or a perfect circle, much depends on the steadiness and practice of the hand if it be done by aim of hand only, but if with the aid of rule or compass, little or nothing; so it is exactly with my plan.

Francis Bacon, *Novum Organum* (1620)

A really good scientist is one who knows how to draw correct conclusions from incorrect assumptions.

Otto Frisch, *What I Remember* (1979)

Logic books are divided into two parts: in the first part on deduction the fallacies are described, and in the second part on induction they are committed.

attributed to Morris R. Cohen

The notion of an effective method, capable of readily solving all problems, is a pervasive one. Both Melanesians and Polynesians when first presented with the awesome power and wealth of the Allied forces in the Second World War could assume only that the white man possessed some special power or method which he had shrewdly kept to himself. As much of their world was governed by ritual and various forms of incantation, they naturally assumed that all the power they could see so plainly before them derived from the exercise of some unknown ritual. Consequently they began to observe, to speculate, and to experiment in the hope that they would discover the elusive ritual whose exercise would bring them the tins, knives, and cloth possessed in such abundance by the whites. Inevitably, they tended to concentrate on the more superficial of white man's rituals. Hat-wearing, marching in formation, stringing wires between trees, whitewashing everything that did not move, and passing chits between each other, were all tried.[1]

Belief in all-powerful methods also emerge, from time to time, in the West. The *Ars Magna* of Ramon Lull, for example, in the thirteenth century had such universal pretensions; as had such later works as John Dee's *Monas Hieroglyphica*, and Leibniz's *De Arte Combinatoria*.[2] Nor, as with the scientologists, are such delusions absent from contemporary society.

The method of science offers less universal claims. Indeed one of the first things a scientist is likely to learn is that there are very definite limits to his powers. There are no perpetual motion machines; the speed of light is constant; and there is no greatest prime. Consequently, methodologists have tended to scorn all universalist pretensions. Bacon, for example, dismissed Lull's work as 'a method of imposture', suitable only for 'meddling wits'.

Science has many techniques, but only one method. Accounts of the lives of scientists often describe how they gained their more specialised skills. For example, Jim Watson has recalled how, as a young man, he was sent from Indiana to Copenhagen to learn some biochemistry, and from there went to Cambridge to learn the techniques of X-ray diffraction analysis. A generation earlier J. D. Bernal had had to leave Cambridge to learn crystallography from the Braggs at the Royal Institution. And so on, throughout the world of science. A surgeon keen to learn a new operation, a physicist wishing to master a new instrument, or a biologist anxious to learn how to culture fragile cells, will all probably be sent to distant laboratories to acquire their novel skills.

Nothing comparable seems to happen in learning scientific method. Promising students are not sent to the seminars of a Carnap or a Popper to master the rudiments of methodology. The very most they can expect would be an ancillary course provided by an adjacent philosophy department. Method as such they are presumably expected either to arrive with or to absorb spontaneously as they go along.

Yet science is a rational activity, one advanced by argument and the presentation of evidence. And since the time of Aristotle the nature and classification of argument has been a central topic of philosophy. Consequently the nature of scientific argument was placed on the philosophical agenda quite early on and has remained there ever since. The initial problem is to identify and describe the precise nature of this supposedly unique scientific method. One answer, long favoured by philosophers and scientists alike, insists that science is, above all else, an inductive discipline (it involves reasoning from particular cases to general conclusions).

Inductivism

Aristotle

The roots of inductive logic can be found in Aristotle. In his *Prior Analytics* Aristotle offered a remarkably comprehensive analysis of one form of argument, the syllogism. This is a purely deductive form of argument and in its most familiar incarnation,

> All men are mortal.
> Socrates is a man.
> Therefore, Socrates is mortal.

there is clearly displayed the essential feature of all deductive arguments, namely, that it is logically impossible to affirm the premises while denying the conclusion.

Yet there are other forms of argument, as Aristotle explicitly noted in his *Topics*:

We must distinguish how many species there are of dialectical arguments. There is, on the one hand, induction (epagoge), on the other reasoning (syllogismos) . . . Induction is a passage from individuals to universals. For example, the argument that supposing the skilled pilot is the most effective, and likewise the skilled charioteer, then in general the skilled man is the best at his particular task. (105.a. 10–19)

As the tradition of inductive logic developed it became more common to describe it as reasoning from the particular to the general, or from the part to the whole. Typical examples found in logic textbooks include:

Gold, copper, lead, etc, are ductile.
Gold, copper, lead, etc, are metals.
Therefore, all metals are ductile.[3]

The most obvious feature of the argument is its invalidity. There is no difficulty in supposing that while gold, copper, and lead are indeed metals and ductile, a metal will be discovered tomorrow which cannot be drawn at all. Precisely this fallacy was committed by the eighteenth-century chemist Antoine Lavoisier in his account of acids. In his *Traité élémentaire de Chemie* (1789) he had claimed that all acids contain oxygen. A great number of combustible bodies, he noted, could be converted into acids by the addition of oxygen. As the union of oxygen with phosphorus, charcoal, sulphur, and saltpetre led respectively to the formation of phosphoric acid, carbonic acid, sulphuric acid, and nitric acid, Lavoisier concluded that oxygen was an element common to all acids.[4]

There was one slight problem. He had been unable, he confessed, to decompose 'the acid of sea salt', now known as hydrochloric acid, but then referred to as muriatic acid. Yet he lacked 'the smallest doubt that it, like all other acids, is composed by the union of oxygen with an acidifiable base'.[5] Lavoisier's confidence in the power of induction and in his misnamed oxymuriatic acid was soon proved groundless. In 1810 Humphry Davy demonstrated that chlorine was an element, and consequently oxymuriatic acid could only be decomposed into chlorine and hydrogen.[6]

Numerous other examples of the failure of what came to be referred to as enumerative induction could be provided, and indeed are to be found in profusion in most textbooks. Clearly, if the method of science was some form of enumerative induction, many of its claims could be held only tentatively. The generalisations of science would forever be hostage to the unexamined instance. Like Lavoisier's oxymuriatic acid and the black swans of Australia, generalisations could easily overthrow our firmly held beliefs. If induction was to survive it would clearly need to be recast.

Bacon

So much had been realised by Bacon when, in his *The Advancement of Learning* (1605), and the self-consciously named *Novum Organum*

(1620), he set out to establish a new approach to science. He began by exposing the limitations of the traditional account of induction. To reach a conclusion merely upon 'an enumeration of instances', he argued, was 'no conclusion, but a conjecture'. In his defence he quoted the case of Samuel searching for a new king among the children of Jesse (I Samuel 16: 1–13). Seven sons were rejected, and if Samuel had reasoned inductively, so too should have been the eighth son, who happened to be David.[7]

To overcome such problems Bacon introduced the refinement known as eliminative induction. Unlike the puerile induction which proceeds by simple enumeration, and which is forever hostage to one contradictory instance, the new form would supposedly 'separate nature by proper rejections and exclusions'. The essential notion behind the technique is that information could be so displayed as to eliminate all false hypotheses, leaving clearly exhibited and in splendid isolation the one remaining true claim. If Abel could have been murdered only by Adam, Eve, or Cain, and if also both Adam and Eve had cast-iron alibis, it must follow that Cain was the guilty felon.

Bacon's own example, presented in Book 2 of his *Novum Organum*, was concerned with the nature of heat. He began by compiling a list of the materials to be studied. This was then divided into tables. In the first, the Table of Essence and Presence, positive instances such as the rays of the sun, fire, warm springs, and quicklime mixed with water, were listed. The second list, the Table of Deviation and Absence, included negative instances among which were moonlight, cold winds, St Elmo's fire, ashes mixed with water, and winter. Using this material Bacon went on to eliminate a number of hypotheses. Because heat is produced by the rays of the sun, it could not be a purely terrestrial phenomenon. Nor could heat be identified with light. This hypothesis could be eliminated by considering the case of the moonbeams which delivered light but no heat to the earth.

After more of this, Bacon finally concluded, for no very obvious reason, 'the nature of which Heat is a particular case appears to be Motion.' There were many things he could point to in evidence. Flames are always in motion, as are boiling liquids, while any 'compression, which checks and stops the motion' also extinguishes the fire. The reasons offered were plausible, and the conclusion sound. Despite this it is difficult to see how it could have been established using the techniques of eliminative induction. The elimination of any number of hypotheses is no guarantee that the next hypothesis to be considered will be acceptable. The method will only work if we have some reason to believe that the original list of hypotheses or evidence was in some sense complete. Such conditions, however, are likely to be satisfied only in the artificial examples constructed in textbooks. A quick check of a car failing to start – battery, spark plugs, petrol etc. – may well rapidly and surely reveal the defect to lie in the carburettor. Nature, however, is often more complex, and certainly more unpredictable, than the workings of the internal combustion engine.

The point can be illustrated by considering the numerous attempts made by the leading nineteenth-century physicist Lord Kelvin to estimate the age of the earth. Textbooks of the period, when discussing energy, would frequently begin by listing the sources of all forms, both kinetic and potential. Taking *all* these factors into consideration, and knowing the rate at which bodies like the earth cool, Kelvin calculated that the earth's surface consolidated about ten million years ago.[8] If we were to go back a hundred million years, then, physicists warned, even assuming the earth to have existed, its surface 'would undoubtedly have been liquid and at a high white heat . . . utterly incompatible with the existence of life of any kind'. And, if geologists objected that this was insufficient time to account for the fossil record, 'So much the worse for geology', the physicists replied. 'Physical considerations . . . render it utterly impossible that more than ten or fifteen millions of years can be granted', they emphasised.[9]

Yet, just as the argument concerning Cain's guilt depends on being able to identify *all* suspects, Kelvin's argument makes the comparable assumption that *all* significant forms of energy had been considered. But, as is well known, Kelvin and his colleagues were totally unaware of the phenomena of radioactivity. First identified in uranium by Becquerel in 1896, it soon became apparent that heat supplied by radioactive transformations in the rocks of the earth would make nonsense of Kelvin's careful calculations. In his 1904 Bakerian lecture Rutherford pointed out that 1 part by weight of radium in 22 million million parts of earth would generate as much heat as was lost by conduction. Clearly the consensus about the earth's age was in need of revision, and within a short time geophysicists had begun to speak of an earth history of more than a billion years.[10]

J. S. Mill

The weakness of the Aristotelian and Baconian approach to induction is evident and consists in, for no apparent reason, restricting scientific inference to a single rule. The aim should surely be to develop as rich a set of rules as possible. There is no reason why scientists should not tackle problems with the aid of enumerative, eliminative, and any other inductive mode which appears useful. In this way the errors which satisfy one rule could well violate another.

One of the best-known modern examples of this approach was presented by John Stuart Mill in his *System of Logic* (1843). For Mill the only way to discover any non-self-evident truth was by induction, and at the root of induction, he argued, there lay the notion of cause.

A man dies, a bridge collapses, vines wither in perfect weather, a current flows through a wire when a magnet rotates, and a star in a distant galaxy explodes. All have causes and the task of the scientist is to discover those causes among the complex of the numerous antecedent factors. Just as there are rules for solving quadratic equations, and rules to determine the direction a current flows through a wire, there should also be, it seemed reasonable to suppose, rules for the detection of causes.

The approach itself does not begin with Mill. The medieval schoolmen, for example, spoke of methods of resolution and composition by which complex phenomena would be first resolved into their composite elements, from which the appropriate cause could be identified. So Robert Grosseteste (1168–1253), in pondering the question why some animals have horns, clearly had in mind what Mill later called the Method of Agreement. All animals with horns, Grosseteste noted, lack teeth in the upper mandible. Unable to defend themselves with their teeth, they developed horns. Much, clearly, remained to be worked out.[11]

The appeal to rules, however, became more refined and also more self-consciously formulated. Hume, in the *Treatise*, devoted two pages to formulating eight rules 'by which to judge of cause and effects'.[12] Most self-conscious of all was Mill with his 'four experimental methods'.[13] These methods can be set out schematically.

1. Method of Agreement

Case	Antecedent events	Effects
1.	ABCDEF	xy
2.	AGHI	xz
3.	ABGHIJ	xzw

As there is only one antecedent event, A, found in all occurrences of x, A is the probable cause of x. For example, the disease thrombo-angitis obliterans (x) produces severe inflammation and thrombosis of the blood vessels of the lower limbs requiring, in some cases, their amputation. The complaint afflicts only smokers (A).

2. Method of Difference

Case	Antecedent events	Effects
1.	AB	x
2.	B	–

Clearly A is a necessary condition for x. Is malaria caused by the foul air found around marshes? Those who live on marshland (B) and are bitten by mosquitoes (A) do develop malaria. But if they continue to reside in their marshland home while defending themselves against mosquito-bites with protective netting, they will not contract malaria.

3. Method of Concomitant Variation

Case	Antecedent events	Effects
1.	ABC	xyz
2.	A + BC	x + yz
3.	A − BC	x − yz

As x varies whenever A varies, A and x are causally related. To a generation brought up on lung-cancer statistics the point is only too clear. The more and the longer we smoke, the greater the risk of lung-cancer, while the less and the more briefly we smoke, the smaller our risk.

4. Method of Residues

Case	Antecedent events	Effects
1.	ABC	xyz
2.	B	y
3.	C	

It is deduced that the remaining factor, A, is the cause of the remaining effect, x. A needle-sharing intravenous drug user (A) who also smokes (B) and drinks heavily (C) develops AIDS (x), cirrhosis (z), and lung-cancer (y). But teetotal smokers with no drug habit develop only lung-cancer, and alcoholic non-smokers who also shun needles develop only cirrhosis. Therefore, it seems, the residue AIDS is caused by the habit of sharing needles with other drug users.

Clearly, given the right kind of example, and the field of epidemiology is better than most, Mill's methods begin to look extremely persuasive. Dozens of case histories could be presented, from John Snow's identification in 1849 of contaminated water as the source of the cholera epidemic, to the present realisation that AIDS is a sexually transmitted disease, showing that on matters literally involving life and death Mill's rules have been regularly employed and have often produced significant results.[14]

But, having said this, it must also be admitted that rules of this type work less well in other areas. In some cases they are simply not applicable; elsewhere they can often identify the wrong or inappropriate causes. First, selection can only be made from an established inventory. If our understanding is so limited as not even to have included the true cause on the list then the procedure will obviously fail. Consider, for example, the well-known connection between malaria and marshland. In French it is actually embedded in the language, as the term for malaria is *le paludisme*, a word deriving from the Latin *palus*, marsh. Hoblyn's *Medical Dictionary* of 1878 confidently derives the disease from 'certain effluvia or emanations from marshy ground'. The identification can be readily supported by running through Mill's rules. The agreement between marshland and malaria is evident, as is the method of difference in the absence of malaria on long sea voyages, in mountains, and in other areas some distance from stagnant marshes. Further, by the method of concomitant variation, the amount of malaria in a population will vary with the spread or decline of the neighbouring marshland. And until someone had the insight to add mosquitoes to the list of possible causes no further application of Mill's or any other inductive rules could improve the situation.

The point had been raised against Mill by his severest critic, William Whewell. The rules, he objected, 'take for granted the very thing which is most difficult to discover, the reduction of the phenomena to formulae such as are here presented to us'.[15] That is, by the time mosquitoes had been added to the list of antecedent events, they had probably already been identified as the cause of malaria. The schema of Mill are therefore more correctly seen as useful and informative means of presenting discoveries already made than as a procedure by which as yet unknown causes can actually be found.

Further, it has been argued, a straightforward application of the rules could often lead to wildly inappropriate results. A series of fires at a chemical works investigated by inductive rules would be likely to identify the invariable presence of oxygen. This information would be of little use to the chemical engineer seeking ways to modify the plant in order to prevent further outbreaks of fire. The causes he is seeking are factors he can control and manipulate. A stiff valve, perhaps, an overheated bearing, a careless attendant – these are the kinds of events likely to appear in his report. Valves and bearing can be redesigned, attendants dismissed; oxygen, however, is not normally amenable to controls of this kind.[16]

Paradoxes

A further difficulty facing any inductivist programme is the appearance at a very basic level of an increasing number of paradoxes. Two in particular have proved to be especially significant. The first, the paradox of the ravens, was proposed by Carl Hempel in 1945.[17] Take the proposition 'All ravens are black', and ask how it could be established inductively. What data, that is, would confirm it? The most natural response is that as long as no ravens which were not black were observed, then the observation of as many black ravens as possible would be the best supporting evidence. Assume also the apparently unexceptional principle:

1 If any evidence E confirms a proposition P, and if P is logically equivalent to Q, then E will also confirm Q.

and the logical truth:

2 'Whatever is a raven is black' is logically equivalent to 'Whatever is not black is not a raven.'

It will therefore follow that anything which confirms 'Whatever is not black is not a raven', a red biro for example, will also confirm 'Whatever is black is a raven'. In this manner the observation of green grass, blue sky, a red dress, a white tea-cup, or a yellow canary – all things which are neither ravens nor black – can be taken as confirming instances of the generalisation 'All ravens are black.'

The assumptions needed for the derivation of the paradox are basic and minimal, as they are also in a second paradox proposed by Nelson Goodman in 1955.[18] Blue sapphires and green emeralds offer inductive support for the propositions 'All sapphires are blue', and 'All emeralds are green.' But, Goodman notes, they offer as much support for the propositions 'All sapphires are bleen', and 'All emeralds are grue.' The terms 'bleen' and 'grue' are defined as follows:

3 x is grue if and only if it is green up until the year 2000 and blue thereafter.

4 x is bleen if and only if it is blue up until the year 2000 and green thereafter.

Thus a green emerald is just as much evidence for the claim that all emeralds are green as it is for the claim that all emeralds are grue.

Claims that the predicates 'grue' and 'bleen' are artificial and in some way fundamentally different from the more natural predicates 'green' and 'blue' have proved to be extremely difficult to justify. 'Grue' it has been objected, unlike 'green', is a complex predicate containing an implicit temporal reference. But the perfectly acceptable predicate 'deciduous' also contains an implicit temporal reference. Again, it has been objected that 'grue' and 'bleen' are derivative, dependent for their definition on the more natural terms 'blue' and 'green'. But, Goodman has replied, it would have been

possible to begin with the concepts 'grue' and 'bleen' and to construct
out of them the derivative definitions:

5   x is blue if and only if it is bleen up until the year 2000
     and grue thereafter.

6   x is green if and only if it is grue up until the year 2000
     and bleen thereafter.

Further examples described by Martin Gardner clearly establish
that the complex notion of confirming a hypothesis cannot be
reduced to the simple relations of inductive logic.[19] If a man of 9 ft
10 in. is disovered, he is a clear confirming instance of the proposition
'All men are less than 10 ft tall.' His discovery, however, weakens the
hypothesis. Consider the hypothesis that no card in a deck of playing
cards has blue pips. A dozen cards with black or red pips are exposed,
all confirming the hypothesis. But then a card with green pips is
turned over, and though it too *seems* to confirm the hypothesis, it also
would seem to cast considerable doubt upon its truth. Thus it must
be allowed that 'there are situations in which confirmation makes a
hypothesis less likely.'

The response of many philosophers to such paradoxes has been to
match the behaviour of the logicians early in the century when
presented with a series of equally damaging antinomies derived from
the foundations of mathematics. Bertrand Russell, Frege, Hilbert,
Zermelo, von Neumann and numerous other mathematical logicians
began to develop new and supposedly more solid foundations free
from all antinomies. Such security, however, seldom comes cheaply.
The systems developed became ever more complex and were frequently forced to adopt quite arbitrary and *ad hoc* axioms and
conventions. And then, suddenly, one day in 1931, Kurt Gödel
proved the amazing thesis that mathematics simply could not have
the kind of foundations mathematical logicians had been seeking to
establish.[20]

So far no Gödel has been able to show that the work of inductive
logicians is similarly misguided. Their work, however, has become
increasingly complex and artificial, and increasingly divorced from
the realities of scientific practice. Consequently, a number of philosophers have sought to describe the procedures of science in quite
different ways.

Hume's problem

A further reason for an unwillingness on the part of many philosophers to take induction seriously is the continuing threat posed by Hume's problem.[21] In his *Treatise of Human Nature* Hume considered under what conditions the following inductive inference could be valid:

All hitherto observed A's are B.

Therefore, All A's, future, past, and distant, are B.

It requires, he argued, the truth of the principle that 'Instances of which we have had no experience, must resemble those of which we have had experience, and that the course of nature continues always the same.' But how is this principle to be justified? Not by reason, Hume answered, as it is not itself a logical truth. We have no difficulty in supposing 'a change in the course of nature', that, for example, bread should no longer nourish, or fire warm. Therefore it must be justified by experience. But this is impossible, as the principle that 'nature continues always the same' has been offered as a justification of experience. It cannot, in turn, and without circularity, be itself justified by experience.

Nor, Hume insisted, can the argument be preserved by resorting to talk of probability. The claim that the course of nature will *probably* continue the same is based on experience just as squarely as the unqualified principle, and it too cannot be justified by experience without circularity.

At this point Hume abandoned logic for psychology. No justification for the inference is possible, and our continued support for the principle is based on habit. Methodologists must also choose. They can seek, Hume notwithstanding, for a justification of inductive inferences. Or, equally heroically, they can accept with Hume that the inductive inferences of science are in fact invalid. Or, more simply, they can turn away from induction and insist that scientific inferences are really deductive.

The hypothetico-deductive method

Plato's likely stories

An alternative tradition can be traced back to Plato. In his *Timaeus* he argued that while we could offer accounts of the 'changeless' which were themselves 'changeless', no such guarantee could be given for descriptions of the physical world of change. This had only a secondary reality and prevented us from always being able to render 'a consistent and accurate account'. We should, therefore, be satisfied 'if our account is as likely as any . . . we should not look for anything more than a likely story in such matters.'[22]

That is, when presented with a problem, in astronomy perhaps, we do not aim to present an account of how the heavens really do work. If this was our aim, we would very likely still be seeking a solution. Instead we aim to construct a story within which the initial problem would appear a natural outcome. For example, Greek astronomers were troubled by the appearance of Venus, sometimes very bright, at other times much dimmer. This could only arise, it was argued, if Venus approached closer to the earth at some point in its orbit. Such a view, however, was incompatible with another notion, namely that the planets moved around the central earth in perfectly circular orbits. What likely story would permit Venus to present itself in this manner? Allow the planets to adopt an eccentric path, with the earth displaced somewhat from the actual centre of the planet's orbit; then, as can be seen in figure 1, the periodic changes in the apparent brightness of Venus make perfect sense within the context of our likely story.

The story is likely in the sense that it is indeed plausible, is compatible with other parts of the story, and does manage to account for a particular difficulty. There is, so far, no need to suggest that the story is true. Consequently, if in the future it becomes less likely, or even most unlikely, when called on to account for further problems, the story can readily be dropped for something more likely.

Such a view has persisted in some form throughout the ages. A classic presentation of the case was made, for example, by Osiander in the preface he added to Copernicus's *De revolutionibus* (1543). Copernicus had argued quite unambiguously that the universe was heliocentric, with the earth just one of the many heavenly bodies orbiting the Sun. Called upon to add an anonymous preface, Andreas Osiander, a Lutheran, discounted Copernicus's own views and argued of his central theses: 'for these hypotheses need not be true nor even probable; if they provide a calculus consistent with the observations, that alone is sufficient.'[23]

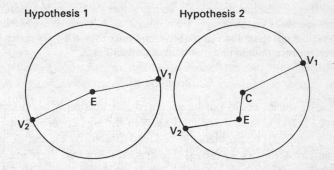

Figure 1.   Two hypotheses about Venus's motion around the earth

Hypothesis 1: Venus moves in a circular orbit around the central earth (E).
Consequence: as Venus is at all times equidistant from earth, it must always
    appear equally bright and equally large.
Fact: Venus at V_2 appears to be much brighter and larger than Venus at V_1.
Conclusion: a new hypothesis is called for.
Hypothesis 2: Venus moves in a circular orbit around a centre (C) some
    distance from the earth (E).
Consequence: Venus at V_2 is closer to the earth than at V_1 and should
    therefore appear brighter and larger at V_2 than at V_1.
Fact: Venus really does appear to be brighter and larger at V_2.
Conclusion: the hypothesis is confirmed.

As Copernicus well knew, this crude form of instrumentalism is
incoherent. The difficulty with likely stories or consistent calculi is
that they tend to exist in quantity. Consequently, the command to
accept a likely story merely invites the response, which likely story?
Copernicus had noted in his own preface to the book that likely
stories had very much got out of hand. It is very tempting, if this style
is adopted, to select a number of likely stories to deal with problems

as they arise, without paying too much attention to how well the various stories form a coherent whole. Precisely this defect, Copernicus complained, was only too evident in the work of his predecessors:

> With them it is as though an artist were to gather the hands, feet, head, and other members for his images from diverse models, each part excellently drawn, but not related to a single body, and since they in no way match each other, the result would be a monster rather than a man.[24]

Cartesian hypotheses

Clearly, stories could not be accepted as true simply because they were likely. Further, we needed a way to choose between competing likely stories. The problems were faced directly by Descartes in his *Principles of Philosophy* (1644). He argued there that there are 'corporeal particles which cannot be perceived by the senses', and that it was in terms of these corpuscles that much of the natural world could be explained. But what of the corpuscles themselves? At first, he seemed to suggest, it was enough 'to explain their possible nature, even though their actual nature may be different'. But he then continued:

> Just as the same craftsman could make two clocks which tell the time equally well and look completely alike from the outside but have completely different assemblies of wheels inside, so the supreme craftsman of the real world could have produced all that we see in several different ways.[25]

Some were, perhaps, satisfied with providing in this manner one out of a number of possible alternatives. Descartes, however, clearly wanted more from his work, and went on to insist that his account of nature was not just likely but 'morally certain'.

Wherein lay the grounds of this moral certainty in the Cartesian position? Descartes replied in terms of a cypher analogy. Suppose someone is presented with a letter encoded in Latin, 'and he guesses that the letter B should be read whenever a appears, and C when b appears . . . If, by using this key, he can make up Latin words from the letters, he will be in no doubt that the true meaning of the letter is contained in these words.'[26] Thus, even though 'his knowledge is based merely on conjecture', it is scarcely credible that there could be alternative solutions. In the same manner, Descartes concluded, one could see that his principles had been able to account for 'all the many properties relating to magnetism, fire and the fabric of the entire world'. It would surely not have been possible 'for so many items to fit into a coherent pattern if the original principles had been false'.

Descartes's new criterion of explanatory power at last offered some independent support for the method. It remained, however, a very limited support. As presented by Descartes it applied only to very general hypotheses which succeeded in explaining many natural phenomena. What, though, if all I sought to explain, or succeeded in explaining, was some particular phenomenon such as the refraction of light? If my hypothesis seemed to work well, but failed to account for magnetism and heat also, it clearly could not benefit from the Cartesian criterion. And, of course, it has to be emphasised that most hypotheses generated in science are of this more restricted form.

The problem of hypothesis selection, therefore, still required attention long after Descartes. Further, after Newton and his apparent rejection of hypotheses, many scientists and philosophers began to argue that the hypothetical method should not be pursued. Philosophers, Benjamin Martin noted in 1769, hold hypotheses in 'vile esteem'. The case against them was most forcefully made by Thomas Reid, the Scottish common-sense philosopher, and his followers:

> If a 1000 of the greatest wits that ever the world produced were, without any previous knowledge of anatomy, to sit down and contrive how, and by what internal organs, the various functions of the human body are carried on . . . they would not in a 1000 years, hit upon anything like the truth.[27]

Further, how could a rational choice ever be made between competing hypotheses? One suggestion was that the simpler hypothesis should be preferred. But, Reid argued, the love of simplicity could often mislead. The world, he proposed, 'is not so simple as the great Descartes determined it to be; nay it is not so simple as the greater Newton modestly conjectured it to be. Both were misled by analogy and the love of simplicity.'[28]

Nor, Reid added, was it satisfactory to point out that the hypothesis permitted the derivation of a number of true consequences. False theories, as well as true, he noted, allow such deductions: 'There never was an hypothesis invented by an ingenious man which has not this [kind of] evidence in its favour. The vortices of Descartes, the sylphs and gnomes of Mr Pope, serve to account for a great variety of phenomena.'[29] Thus, while it was essential that hypotheses could account for all the evidence, this accounting alone was insufficient to prove the truth of the theory.

Prediction

Clearly, something further was required. The final element emerged in the nineteenth century and has remained an essential feature of the hypothetical methodology ever since. To transform a hypothesis which merely explained what was already known into something more authoritative, it would be necessary to show that the hypothesis could be used to explain or predict phenomena it was not specifically designed to explain or predict. The more unexpected or otherwise 'unpredictable' these phenomena would previously have been, the more credible the hypothesis. For example, two or more theories of light could be produced to account for the familiar phenomena of reflection and refraction. At this level there could well be nothing to choose between the competing hypotheses. Such in fact was the fate of the wave and corpuscular theories of light proposed by Huygens and Newton in the late seventeenth century.

In 1816 Fresnel submitted a *Mémoire* to the Paris Académie on how the wave theory could best account for the diffraction of light. The mathematician, Simeon-Denis Poisson, noted an unexpected, indeed startling consequence of Fresnel's theory: at the centre of a shadow formed by a circular disc there should be a bright spot. The prediction was quickly confirmed, providing at the same time very strong confirmation of Fresnel's hypothesis.[30]

At last a procedure had been introduced which offered hypotheses some kind of independent support. Much of eighteenth-century science had been concerned with the variety of imponderables – caloric, the fluid of heat, magnetic effluvia, a luminiferous ether, a gravitational ether – in terms of which most of the physical world was explained. Against criticism of this approach, the powerful explanatory role of the imponderabilia could always be deployed. But, at last, a more pertinent question could be asked. Magnetic effluvia, it could be conceded, may well explain all known magnetic phenomena. But did it also predict the existence of any new and unexpected phenomena? And had the existence of these new phenomena actually been confirmed? Once it was demanded that hypotheses must have predictive powers much of the furniture of eighteenth-century science began to look to more critical minds decidedly speculative.

The issue itself was worked out by, among others, William Whewell in his *Philosophy of the Inductive Sciences* (1840). Hypotheses should, he noted, be consistent with '*all* the observed facts'. But, further, they should successfully foretell 'phenomena which have not yet been observed'. To illustrate his thesis he turned to advances recently made in the field of optics. The corpuscular theory of light had been able to explain 'the phenomena which it was at first contrived to meet'. But only the wave theory had been 'extended to facts not at first included in its domain'.[31]

A challenge to Whewell's account of scientific method came almost at once from John Stuart Mill. 'Predictions and their fulfilment' were, he argued, only fit 'to impress the uninformed'. Mill did not deny that successful predictions were welcome; he claimed merely that they provided no *extra* support for the initial hypothesis.[32] The point can be illustrated with the aid of a familiar example. In 1915 Einstein published his account of general relativity. It carried the unexpected implication that rays of light grazing the sun should be deflected 1.7 seconds of arc. The observation could only be made during a total solar eclipse. Fortunately such an event was due in 1919, and consequently in Principe in May 1919 Eddington was able to make the observations which confirmed Einstein's claims. The story is often presented as one of the more telling examples of the power of the hypothetico-deductive method in science.

It could, however, have worked out differently. In 1914 the German astronomer Erwin Finlay-Freundlich had planned to measure the deflection, expected to be just 0.83 seconds of arc, during an eclipse visible in Russia. The outbreak of war stopped all such plans. Presumably, if Finlay-Freundlich had been allowed to observe the eclipse he would have found the deflection to have been the 1.73 Eddington observed in Principe in 1919.

Thus, when Einstein came to publish his 1915 paper he would merely have been able to show that his theory could account for an *already* established result. Would this really have been any different from what actually happened?, Mill would have asked. The theory remains the same, as do its predictions and confirmations. Surely, it is a matter of drama or rhetoric that distinguishes the two scenarios, not logic. Whether the observations were made post-theory or ante-theory is irrelevant.

Hoffmann, Einstein's biographer, has argued that there *is* a difference. After Finlay-Freundlich's '1914 observations', Einstein's 1915 calculations would have appeared 'tame': 'People would have felt that he had made an arbitrary ad hoc adjustment – which in fact he had not – and the deflection of light would have lost the tremendous impact that it had as a prediction.'[33] Yet Hoffmann's instructive tale, expressed so openly in psychological language, would have been unlikely to have converted Mill. It is always possible to increase the impact of a speculative theory in molecular biology by suggesting that it could well lead to a cure for cancer. The ploy may well win headlines; it does little to justify the theory. There remain, however, more important objections against this account of scientific method.

Objections to the hypothetico-deductive approach

The fundamental questions still remain. Is the hypothetico-deductive approach an accurate account of the procedures of science? And, if so, is it a legitimate method of reasoning? Many contemporary scientists, Peter Medawar for example, have agreed that it does describe much of their own work. Nor is it a difficult task to cast many of the better-known and more successful scientific theories in this mould. Yet doubts remain whether these accounts are closer to imaginative reconstructions than accurate descriptions.

Consider the case of Charles Darwin and the fate of evolutionary theory. It could be said that Darwin proposed a hypothesis, that of natural selection, and that it permitted a number of successful predictions. The truth is rather different. Darwin published his views in 1859 in his *Origin of Species*. Before his death in 1882 the *Origin* went through a further five, often substantially revised, editions. They were revised so frequently and so radically because Darwin had found it increasingly difficult to deal with the problems presented to him by some of his more acute critics.

His main difficulty was to explain how the units, the variable traits, that natural selection worked on, could be inherited. Critics began to draw consequences from his theory, and concluded that they could not be realised in Darwinian terms. Furthermore Darwin himself came to see the force of his critics' case and his own inability to offer a satisfactory reply. The problem was to work out how favourable variations could ever survive long enough in a population to be selected. Prevailing theories of inheritance suggested that they would be diluted and disappear very quickly from the general population. Darwin responded with a new theory of inheritance, pangenesis, which was quickly put to the test and found to be undetectable.[34]

At this point Darwin, according to the rules of the prevailing methodology, should have acknowledged the weakness of his position and abandoned his theory. Instead, he offered the weakest of responses, an excuse commonly offered in defence by the fake and the charlatan.

One critic, the engineer Fleeming Jenkin, had objected that there appeared to be very definite limits to the rate of variation found in nature:

Even though it were only one part in a thousand per twenty or thirty thousand generations, yet if it were constant or erratic we might believe that, in untold time, it would lead to untold distance; but if in every case we find that deviation from an average individual can be rapidly effected at first, and that the rate of deviation steadily diminishes till it reaches an almost imperceptible amount, then we are entitled to assume a limit to the possible deviation.[35]

Darwin did not deny the facts of the argument. He replied with the claim that 'It would be equally rash to assert that characters now increased to their utmost limit, could not, after remaining fixed for many centuries, again vary under new conditions of life.'[36]

Given the right to call in the outcome of 'many centuries', almost anything would be possible. The alchemist could equally claim his powder would transform lead into gold once it had been allowed to stand for a few further centuries. There could even be a lazarus ointment which when smeared on a corpse would revive it several centuries in the future. Clearly, Darwin's argument, one frequently deployed, is very weak.

Yet it is one commonly met with, in some form, in science. A further example from more recent times has been recorded by Richard Feynman. At one time he worked with Murray Gell-Mann on the theory of beta decay, the process by which a nucleus transforms itself by emitting an electron or positron. Although Feynman thought the theory 'rather neat', it predicted that electrons should be emitted from a disintegrating neutron equally in all directions. But experiment showed that '11% more came out in one direction than the others'. Feynman did not dispute the results. The experimenter, Telegdi, he noted 'was an excellent experimenter and very careful'. When asked by Gell-Mann what they should do with their work, Feynman replied: 'We just wait.' The wait was extremely short. Two days later Telegdi reported an experimental error. His work was now in 'complete agreement' with the new theory of beta decay.[37]

Darwin had to wait much longer before the confusions in the objections to his work from Jenkin and others could be finally exposed. It is now clear that Darwin lacked the material to dispose of the objections himself. It would require the discoveries of Mendel and others before it could be shown that the kind of variation required by Darwin could in fact be inherited. Darwin was therefore quite right to insist on the value of his work despite the unfavourable predictions it led to.

The above examples lead to an alternative version of the hypothetico-deductive method. The method of science consists in taking a hypothesis such as natural selection and pursuing its implications and consequences wherever they lead, *despite* any unfavourable predictions or explanatory failures. It could even be said that it is precisely this feature which identifies the creative scientist. Obsessively he will seize upon an idea and develop it against all opposition, and against all counter-evidence, for the rest of his life. The actual framing of the idea is a relatively trivial matter. More often than not it will be found that the same idea was considered by many of his contemporaries. But they, impressed by the difficulties involved, and acutely aware of unsuccessful predictions, rapidly abandoned the project for less intractable problems.

If there is one feature shared by such creative figures as Copernicus, Kepler, Newton, Lyell, Dalton, Darwin, and Einstein, in addition to their formidable intelligence, it is an enormously strong and single-minded will. Whether this is seen as stubbornness, or even obsessiveness, is irrelevant. In all these cases, however, it is apparent in the manner in which an important hypothesis was adopted and pursued, come what may, throughout their entire lives.

A further criticism of the hypothetico-deductive approach is that it can very easily lead to the adoption of false theories. This can be seen in the case of sympathetic powder, one of the more curious aspects of seventeenth-century science. Paracelsus was supposed to have developed a salve which would heal at a distance. Smear the salve on a bandage, or indeed the weapon which had caused the wound, and the virtues present in the salve would be attracted to the wound by a kind of magnetic sympathy and heal the wound. The notion was picked up by Sir Kenelm Digby, a founder member of the Royal Society, and described in his *A Late Discourse . . . Touching the Cure of Wounds by the Powder of Sympathy* (1658), a work published in at least twenty-five further editions before the end of the century.[38]

Patients treated with sympathetic powder fared better than those who had their wounds dressed repeatedly by the contaminated hands of a nurse or physician. Left to themselves most wounds will heal. The advantage of sympathetic powder was that, as it claimed to heal at a distance, the physician was discouraged from actually infecting the wound himself. Consequently, weapon ointment would indeed appear more efficacious than other methods of treatment and the hypothesis of the existence of potent magnetic sympathies would thereby have been established.

The trouble is, basically, that a false hypothesis can lead to surprising and true predictions just as readily as a true hypothesis. Something of the complications in nature which lead to this kind of result can be seen in the following example. Early in the present century cot deaths were often attributed to an enlarged thymus gland. An article in the *British Medical Journal* of 1905 noted that 'Enlargement of the gland is a recognised cause of sudden death in infants, owing either to direct pressure on the trachea, or to its pressure on the vagus nerve causing spasm of glottis.'[39] It would follow, assuming the correctness of the hypothesis, that in other deaths the thymus would be a normal size. The prediction was duly confirmed and 'the theory of death by thymus enlargement continued to hold sway for more than 50 years.'

But, it turned out, the pathologists were mistaken. The enlarged thymus they had observed in the victims of cot deaths was in fact the normal, healthy state. The confusion had arisen because:

The only autopsy specimens available to the pathologists at the time were those from deaths due to malnutrition or infection, in both of which the thymus becomes involuted and small. Thus, the apparently enlarged thymus in children dying from overlying was in fact the normal gland.[40]

Luck and originality

Method clearly can have no more than a limited use within science, if only because important discoveries can be made in quite unmethodical ways. There is first the role of accident or luck in science. Admittedly the accidental observation, the lucky malfunction, will only strike the attuned mind, but they remain matters of chance and accident none the less. And without them well-designed research can often prove to be totally unrewarding.

One area rich in serendipity is chemotherapy. Frequently, drugs are discovered by accident, and when discovered often turn out to have unexpected properties. Chlorpromazine, the major tranquilliser, for example, was first used to treat worms, and iproniazid was found to have antidepressant properties while being tested for efficacy against TB. Or, consider the case of disulfiram, also known as Antabuse (meant to stop one drinking), which produces exceedingly unpleasant effects when taken with alcohol. In 1948 Danish pharmacologists were testing tetraethylthiuram disulphide as yet another treatment for worms. Although it worked well with rabbits it produced most uncomfortable symptoms when the Danish workers took the drug themselves and subsequently had a drink.

The best-known example in this area, however, remains Fleming's discovery of the therapeutic value of penicillin. In September 1928 Fleming returned from a one-month holiday to a laboratory workbench stacked with culture dishes awaiting his examination. On one of the plates, he noted, 'around a large colony of contaminating mould, staphylococcal colonies became transparent and were undergoing lysis'. Fleming's casual observation led little more than a decade later to the development of penicillin as the first antibiotic.

The events leading to Fleming's discovery have been reconstructed by his former colleague Ronald Hare and found by him to be so unusual that it is doubtful whether they can have occurred 'more than a very few times since bacteriologists first started to cultivate microbes on culture plates'.[41] It would have been necessary, Hare has shown, for the temperature not to have risen above 68°F for five consecutive days, for the mould to have contaminated the dish within a few hours of seeding the staphylococci, for Fleming to have set aside his cultures to await his return, and for there to have been an available source of the appropriate penicillin. An examination of meteorological and other records convinced Hare that these conditions would have been met in Fleming's laboratory during the first week of July 1928. He concludes that the discovery of penicillin is, surely, 'the supreme example in all scientific history of the part that luck may play in the advancement of knowledge'.[42]

Many other examples are known. Beveridge actually lists some nineteen examples 'in which chance played a part' in an appendix to his *The Art of Scientific Investigation* (1957), in addition to devoting a whole chapter to the topic. A typical example from Beveridge concerns the discovery by A. V. Nalbandov in 1940 of a way to keep chickens alive after the surgical removal of their pituitary. At first all Nalbandov's chickens died shortly after the operation.[43] Then, for no apparent reason, 98 per cent of a later group survived without any difficulty. The result was not repeated, and in further groups all the birds died. Many factors were considered but were soon found to be irrelevant. The answer finally came when Nalbandov happened to drive past his laboratory at 2 a.m. one morning and noted that the 'lights were burning in the animal room.' A substitute janitor was on duty, one who customarily kept the lights on all night. He had also been on duty during the days of Nalbandov's earlier 98 per cent success rate. At this point orthodox methodology could begin to operate. In a short time Nalbandov had worked out that chickens without a pituitary do not eat in the dark and consequently develop a fatal hypoglycaemia. In contrast, chickens provided with a few hours' nocturnal light ate normally and consequently survived the trauma of their operation.

There is also the question of originality and novelty. Originality itself is difficult to describe realistically. The great triumphs of the past are by now so familiar, the subjects of so many studies, that they appear more a routine matter than the highly startling proposals they once were. Alfred Wegener's theory of continental drift, for example, has long since lost its power to shock. Yet many of his contemporaries found his proposals not merely improbable, but outrageously speculative. This was no exercise of scientific method, his critics complained; he had simply framed an initial idea and then made 'a selective search through the literature for corroborative evidence'.[44] Facts which did not fit in with his views were simply ignored. His critics were right: there is no method, scientific or otherwise, to account for Wegener's insights. According to one story the idea came to him as he observed ice splitting and separating in the Greenland seas. In other accounts he was impressed by the apparent congruence between the Atlantic coastlines of Latin America and Africa. In either case his work seems to have been based on casual observations which must have been made on numerous occasions by most of his professional colleagues.

At this point Wegener, as his critics claimed, did little more than assemble whatever evidence he could find in favour of his hypothesis. Although there is no method whose exercise would lead to the framing of a hypothesis of this kind, there are many rules against which the hypothesis, once formulated, could be tested. So the geophysicist Harold Jeffreys in his influential *The Earth* (1924) could examine the forces proposed to account for continental drift and estimate their strength. None of them, he concluded, were powerful enough to overcome the earth's strength.[45] Against such authoritative criticism Wegener could do little more than offer the familiar defence: 'The Newton of drift theory has not yet appeared.'[46] Until his appearance we would have to wait for the solution to the problem of what forces drive the continents away from each other.

Deductivism in practice

In their discussions of scientific method philosophers normally concern themselves with general issues and seldom bother to show how scientific results are actually deduced. A notable exception to this indifference can be found in the work of Rudolf Carnap. Consider, for example, what happens when a logician like Carnap sets out to establish what happens when iron is heated.[47]

He begins with the axioms:

A1. For every x, t_1, t_2, l_1, l_2, T_1, T_2, a, [if (x is a Sol and lg $(x,t_1) = l_1$ and $lg(x,t_2) = l_2$ and $te(x,t_1) = T_1$ and $te(x,T_2) = T_2$ and $th(x) = a$), then $l_2 = l_1(1 + a(T_2 - T_1))$],

where t, l, and T stand respectively for time, length, and temperature, Sol = solid, te = temperature of . . . , th = coefficient of . . . , lg = length of . . . , a = coefficient of thermal expansion, and x a piece of iron.

A2. For every x, if [x is a Sol and x is a Fe], then the $th(x) =$
0.000012,

where Fe = iron.

Thus, Axiom 1 simply states that any change in length of a heated
solid is a function of its coefficient of thermal expansion, its original
length, and the temperature applied. Axiom 2 merely adds the fact
that the coefficient for solid iron is 0.000012. Also required are some
mathematical theorems of the kind:

T1. $1000 \times 0.000012 \times (350 - 300) = 0.6$,

and some initial assumptions:

As1. c is a Sol
As2. c is Fe
As3. $te(c,0) = 300$
As4. $te(c,600) = 350$
As5. $lg(c,0) = 1000$.

From all of which it is a relatively simple matter to infer that the 50°
increase in temperature was correlated with an increase in a 1000 cm
iron bar of 0.6 cm. The inference may be simple; it is also exceedingly
tedious and lengthy. And this is, of course, just one of the hundred or
more inferences a scientist might be expected to make in a day's
work.

Conclusion

Despite being very much in favour with a number of distinguished
scientists, there are sound reasons for concluding that the hypothetico-deductive method does not accurately describe the methodology of science. Applications of the method can readily lead to false
theories, and important and influential theories can be constructed
beyond its confines. Further, the system allows little room for
creativity, originality, or even luck. It is all very like the attempts
sometimes made by successful novelists and directors of courses on
creative writing to lay down rules for the writing of novels. Yet it still
happens that novels which scrupulously follow the perceived rules
can be totally unreadable, and others which violate every maxim in
sight can enthrall. Science, no less than painting, cannot be done by
numbers.

7

CAN PHILOSOPHY LEARN FROM SCIENCE?

> What I wish to bring to your notice is the possibility and importance of applying to philosophical problems certain broad principles of method which have been found successful in the study of scientific questions.
>
> Bertrand Russell, *On Scientific Method in Philosophy* (1914)

> Philosophers consistently see the method of science before their eyes, and are irresistibly tempted to ask and answer questions in the way science does. This tendency . . . leads the philosopher into complete darkness.
>
> Ludwig Wittgenstein, *The Blue Book* (1933)

What exactly is the method of philosophy? How do philosophers, from the comfort of their armchairs, pursue their researches? With no laboratories to attend, experiments to conduct, or primary sources to seek out, it is far from obvious how they fill their day. To be told that they construct and examine arguments is insufficient, and will fail to distinguish them even from lawyers and politicians. Nor are accounts by philosophers of their own working days any more revealing. Bertrand Russell, for example, has described how he spent the summers of 1903 and 1904 working on his planned *Principia Mathematica*: 'Every morning I would sit down before a blank sheet of paper. Throughout the day . . . I would stare at the blank sheet. Often when evening came it was still empty.'[1]

Many philosophers have, in consequence, spent much of their careers, not in tackling philosophical problems, but in obsessive pursuit of the proper philosophical method. The writings of Plato, Descartes, Kant, and Wittgenstein are saturated with such concern. Others, unsure of their own methodology, have preferred to consider the procedures of more straightforward disciplines. Hence the philosophical attention paid to scientific method. More recently philosophers have turned to the analysis of the methodologies of individual sciences. Consequently, in addition to textbooks on inductive logic, monographs have begun to appear on the methods of such specific disciplines as social anthropology, evolutionary biology, microscopy, and quantum mechanics.[2] Philosophical attempts to describe and evaluate the methods of science will form the subject of the following chapter. But first the more immediate impact of science on philosophical method will be examined.

In the past philosophers, with no agreed method of their own, have tended to be opportunists, borrowing methods and procedures from whatever disciplines appeared more successful. As few disciplines have ever seemed to be as successful as science, and as many philosophers came from a scientific background, they turned naturally to science as a source of reliable methods. In this approach philosophers have done no more than follow the example of scientists themselves. The success of Newtonian physics, for example, as will be seen in greater detail below, convinced many chemists, physiologists, and psychologists that great intellectual revolutions could be produced if only they could work out how to introduce universal attractive forces into their own disciplines.

But not all methods are good travellers, even for distances far shorter than that between physics and psychology. Consider, for example, one of the great successes of nineteenth-century biology, the germ theory of disease. By 1880, after the work of Pasteur and Koch, it had at last become clear that many major, common diseases were spread by bacteria. Within a decade the bacteria responsible for cholera, anthrax, leprosy, gonorrhoea, relapsing fever, and TB had all been identified. The way was clearly open for others to detect the bacteria responsible for the remaining common diseases. As few diseases are as common as malaria, many workers in the 1870s began to search the blood of their malaria patients for any likely germs.[3]

Before long, in 1879, E. Klebs and C. Tommasi-Crudelli reported that they had found in the Italian swamps a bacillus which, when injected into rabbits, produced malaria, and which they named *Bacillus malariae*. Under this name it can be found frequently referred to in the textbooks of the 1880s.[4] But, as is well known, malaria is not caused by bacteria lurking in the swamps of Italy or anywhere else. It is in fact caused by various protozoa, and is transmitted through the bite of certain mosquito species. The notion of an insect vector, however, was unfamiliar to the microbiologists of the 1870s. Dazzled by the success of Pasteur and Koch, and unaware of the complex lifecycle of the malaria parasites, many workers all too readily assumed that the new techniques of microbiology could be applied uncritically to any disease.

Even more striking is the case of the French mathematician and astronomer Leverrier.[5] In 1846 he gained international fame by successfully predicting the existence, mass, and location of the previously unobserved planet Neptune. Leverrier's work was based on known anomalies in the orbit of Uranus. Such behaviour, he argued, could be caused only by a gravitational attraction exerted on Uranus by a massive body. Using the laws of Newtonian mechanics, Leverrier, with considerable difficulty, finally succeeded in working out precisely how massive, and exactly where such a body would have to be. When the astronomer Joseph Galle turned his telescope to the indicated place in the heavens on 23 September 1846, he found the predicted planet.

The success of Leverrier was taken by many as the surest of all signs of the soundness of his methodology. Consequently when Leverrier also noted a persistent and inexplicable perturbation in the orbit of Mercury, he began to suspect the presence of a further unknown planet. Using precisely the same techniques as those which had led to the detection of Neptune, Leverrier concluded in 1859 that there was a planet inside Mercury's orbit responsible for the observed anomalies. He named the planet Vulcan. This time, however, all attempts to observe the new planet proved unsuccessful. And, further, Einstein in 1915 offered an alternative explanation of Mercury's orbit which did not require the existence of a new planet.[6]

If techniques worked out for the study of TB prove initially misleading when applied to malaria, and if Leverrier's extension of his work on Uranus to Mercury has proved to be equally misleading, it must be very uncertain that the transplantation of scientific techniques to the more distant area of philosophy will fare better.

Three methods throughout the long history of science have proved to be especially successful – the deductivism of Euclid, Newton's use of attractive forces, and Darwin's deployment of natural selection to explain biological phenomena. Consequently, it is these three methods that philosophers have most frequently and most notably adapted to their own use. The results of their efforts form the subject of the next section.

Deductivism

The first major successes in science came through mastery of the deductive method. The best example of this approach remains Euclid's *Elements*. Dating from about 300 BC, the *Elements* survived as a working mathematical text until the beginning of the present century; as a model, however, for mathematics and other disciplines, much of its potency has been retained.[7] At the age of eighty-five Bertrand Russell could still recall how as an eleven-year-old boy he had been introduced to Euclid: 'This was one of the great events of my life, as dazzling as first love. I had not imagined there was anything so delicious in the world.'[8]

From the time of antiquity onwards the deductive approach was restricted neither to Euclid, nor to geometry. Euclid himself had also composed an *Optics* in which, from a number of initial postulates, he set out to derive conclusions about light rays with as much rigour as he had derived theorems about triangles from the axioms of the *Elements*. He was followed by Archimedes, who not only used the deductive approach to describe such purely mathematical objects as spheres and conoids, but also sought to show in his *On Floating Bodies* that physical data could be treated in the same manner. The tradition survived in such later works as the *Conics* of Apollonius, Hero's *Catoptrics*, and Ptolemy's *Optics*.[9]

Before proceeding further it would be well to consider precisely wherein lies the strength of deductivism. Three features particularly stand out, revealing the extraordinary and unique power of the approach.

1 The first great benefit of deductivism is absolute certainty. If

we start from self-evident propositions, and use only indisputable principles of inference, then any conclusion reached must be equally indisputable. Thus from such self-evident truths as:

Postulate 4. All right angles are equal to one another,

and such indisputable principles of inference as:

Common notion 2. If equals be added to equals, the wholes are equal,

together with such quite unexceptional definitions as:

Definition 11. An obtuse angle is an angle greater than a right angle,

Euclid went on to deduce such landmarks of western intellectual history as Pythagoras's theorem, that there are an infinite number of primes, and that the square root of 2 is irrational. And these are, of course, propositions whose truth is guaranteed by logic.

2   Further, deductive truths are objective truths. As a deductive proof can be checked in a mechanical manner by anyone familiar with the language, all disputes about their validity should be resoluble. Just as there can be no sensible dispute about the accuracy of the calculation $778 \times 183 = 142,374$, there can be none about the theorems of Euclid. To discover the theorems themselves may well require some degree of mathematical sophistication, but their appraisal is more a matter of routine.

Precisely this feature has been recognised, often with envy, by workers in more contentious areas. Letwin, for example, in his study of the origins of economics, noted the frequency with which the charge of self-interest was levelled against early writers.[10] It was easy enough to discount the pleas of wool merchants, say, for the imposition of tariffs on imported woollen goods. What, however, if the writer was unknown or anonymous? How was it possible to distinguish between proposals which were merely self-serving, and those that, despite appearances, made economic sense? The answer, oddly enough, came from geometry. No one, it was noted, had ever accused Euclid of defending one of his theorems out of self-interest. Economic arguments must therefore be cast in the same mould, with conclusions derived in a 'limpid and intelligible' manner from 'clear and evident truths'. In this manner Roger North, an early writer on economic questions, concluded, his arguments would at least be listened to and not dismissed out of hand as special pleading.

3 If deductivism could offer no more than certainty and objectivity, it would be of little interest. Both features are compatible with total triviality and we could have a system telling us merely that triangles had three sides, squares four sides, and right angles 90°. Consequently, it is vital that deductivism has as a third feature the power to establish original and powerful truths. Occasionally the truths are so novel and strange as to be scarcely credible. Consider, for example, the Banach-Tarski theorem, a truth so counter-intuitive that it could have arisen nowhere other than in the remoter branches of a deductive system. The theorem states that under certain conditions two spheres S_1 and S_2 with different radii can be divided each into the same finite number of disjoint sets, all congruent to each other. As spelt out by Kasner and Newman, the theorem states: 'there is a way of dividing a sphere as large as the sun into separate parts, so that no two parts will have any points in common, and yet, without compressing or distorting any part, the whole sun may at one time be fitted snugly into one's vest pocket.' Alternatively, a pea 'may have its component parts so rearranged that without expansion or distortion . . . they will fill the entire universe.' 'No fantasy of the Arabian nights', Kasner and Newman conclude, 'can match this theorem of hard, mathematical logic.'[11]

Less spectacularly, the deductive approach can render certain and clear what was earlier tentative and suspected. Consider the case of the regular solids. These are solid figures, like the cube, each of whose faces are regular, congruent polygons. The cube, for example, is composed of six congruent squares. How many such figures are there? As there is an infinite number of regular polygons, it might have been thought that a good many regular polyhedra could also be constructed. Craftsmen found, however, that they could make a cube from six squares, a dodecahedron from twelve pentagons, and that four, eight, or twenty equilateral triangles could be combined to form regular tetrahedra, octahedra, and icosahedra respectively. And there the matter rested. Despite all efforts no one could construct a sixth regular solid. Was this a genuine constraint of space? Or merely a lack of imagination shown by designers and artists? Could there not be regular polyhedra with, say, 256 faces, or perhaps even 1,000 faces? Without the power of deductive geometry the issue would still be in doubt. Euclid, however, as the final theorem of the *Elements*, proved in a natural and straightforward manner that 'no other figure, besides the said five figures, can be constructed which is contained by equilateral and equiangular figures equal to one another.'[12]

Cartesian deductivism

Such are the general attractions of deductivism. What, though, of its more specific appeal for philosophers? As modern deductivism starts with Descartes, it would be appropriate to begin with the young French philosopher as, in the 1620s, he surveyed the state of learning. These early views can be seen in his *Rules for the Direction of the Mind*, composed in about 1628. He noted, as have critics from Socrates onwards, that 'hardly anything is said by one writer the contrary of which is not asserted by some other' (Rule 3). Nor did the study of ancient authorities seem fruitful. From them we could learn history, but not science.

More importantly, he stressed, scholars were hopelessly unsystematic in their approach. Chemists, geometers, and philosophers were no better guided than 'someone who is consumed with such a senseless desire to discover treasure that he continually roams the streets to see if he can find any that a passer-by might have dropped' (Rule 4). If they did find some truth, it was the result of fortune rather than diligence. To emphasise the point Rule 4 stated quite simply that 'We need a method if we are to investigate the truth of things.'

It was a time in which many were seeking such a method. In the previous century some, like Cornelius Agrippa, had sought enlightenment from cabbalistic and hermetic techniques, while in Paris the Ramists aimed to develop a new dialectic. They were followed in the seventeenth century by Bacon and his attempt in his *Novum Organum* (1620) to create a new method for science. At the same time, Montaigne, Sanchez and other sceptics were denying the very possibility of any such method.

As presented in his *Discourse on Method* (1637), the Cartesian methodology consisted of four rules. By the first rule Descartes bound himself 'never to accept anything as true if I did not have evident knowledge of its truth . . . and to include nothing more in my judgements than what presented itself to my mind so clearly and so distinctly that I had no occasion to doubt it.' Or, in other words, to accept as premisses only propositions as self-evident as the axioms of geometry. The method of drawing conclusions was described in the third rule: he aimed 'to direct my thoughts in an orderly manner, by beginning with the simplest and most easily known objects in order to ascend little by little, step by step, to knowledge of the most complex.'

Descartes made no secret of the source of his method. After stating the four rules he went on to comment: 'Those long chains composed of very simple and easy reasonings, which geometers customarily use to arrive at their most difficult demonstration, had given me occasion to suppose that all the things which can fall under human knowledge are interconnected in the same way.'[13]

Did Descartes actually carry through his programme and show as a geometer the interconnectedness of things? Even the most casual examination of the Cartesian texts reveals that far from presenting his arguments deductively, he used as discursive a form as any other philosophical author of the period. Critics were quick to pick up this discrepancy between Cartesian propaganda and practice. The point was raised by Mersenne in the published 'Objections' to Descartes's *Meditations*.[14] Why, he asked, had not Descartes, an expert mathematician, developed his system in the manner of Euclid?

Descartes took the point seriously. His answer was based on the classical distinction between analytic and synthetic modes of reasoning first drawn explicitly by Pappus in the fourth century AD.[15] Synthesis was the way of Euclid and the geometers. Starting with first principles, the synthetic thinker drew from them whatever conclusions seemed to follow. If he was fortunate he would find among the many theorems of the system whatever proposition he had hoped to prove. In contrast the analytic thinker began with the proposition he wished to prove and tried to trace its origin back to the first principles the synthetic thinker began with. Descartes offered two reasons to explain why he had not followed the synthetic approach.

To begin with, he argued, it could be confusing. Often, before a particular theorem can be deduced from a set of axioms, a large

number of initial theorems may have to be derived. Textbooks of
mathematical logic are notorious for the elaborate machinery they
must first set up before, usually in the second volume, they succeed in
proving $1 + 1 = 2$.[16] Just as it would be insane to allow mathematical
logicians to teach simple arithmetic in primary schools, it would be
equally confusing to present metaphysics geometrically.

Secondly, while the axioms of geometry are likely to be granted by
all, the assumptions of metaphysics usually command less support.
Indeed, Descartes noted, 'nothing in metaphysics causes more
trouble than making the perception of its primary notions clear and
distinct.'[17] It was therefore wiser, as a matter of prudence if not of
logic, to adopt the analytic approach.

Yet, Descartes insisted, there was no real difficulty in demonstrating the existence of God and the distinction between mind and body
'in geometrical order'. To prove his point he actually provided such a
demonstration in his reply to Mersenne. It is a curiously inept
exercise, and bears no more than a casual resemblance to the
demonstrations of Euclid. Ten definitions, seven postulates, and a
further ten axioms are pressed into service to prove four theorems
and one corollary. Further, in the place of the simple, single-line
axioms of Euclid, those of Descartes run to, in one case, some 150
words. Few are self-evident, and some are not even true. Axiom 8, for
example, states: 'That which can accomplish the greater or more
difficult, can also accomplish the less or the more easy.'[18] Many a
gambler has fallen foul of this principle by rashly assuming that a
horse which wins the Derby cannot be beaten in a lesser race.

In fact Descartes followed neither the analytic nor the synthetic
method with any consistency. He was equally reluctant to observe his
numerous other rules. Consider, for example, his account of light
and colour. The sensation of light, for Descartes, was simply 'the
pressure of the material of the second element'.[19] What, then, of
colour? The particles have both a rectilinear motion and spin. The
rectilinear motion provides the sensation of light, and as for colour:
'If the speed at which they turn is much smaller than that of their
rectilinear motion, the body from which they come appears *blue* to us;
while if the turning speed is much greater than that of their rectilinear
motion, the body appears *red* to us.'[20] Whether light particles rotated
or not was something Descartes could neither deduce from first

principles nor observe. His selection of this property was as arbitrary as the effect he assigned to it. On numerous other occasions he described equally arbitrary mechanisms to account for the familiar phenomena of nature.

It is clear from the Cartesian approach that deductivism is far easier to proclaim than to follow. If taken seriously, propositions can no longer be held because they seem to be plausible, reasonable, or even probable. Only those propositions which can be derived from the initial self-evident axioms can be supported. But constraints of this kind are likely to prove far too restrictive for someone intent on developing a comprehensive account of the physical universe. They are consequently, as by Descartes, simply ignored.

Spinoza

A more serious commitment to deductivism was shown by Spinoza. In an early work published in 1663 he took the trouble to show how Descartes's *Principles of Philosophy* could be 'demonstrated in a geometrical manner'. This was followed with his own *Ethica ordine geometrico demonstrata* (1677). His aim would be, he announced, to consider 'human actions and desires in exactly the same manner' as he considered 'lines, planes and solids'.[21]

No other philosophical work shows such a total commitment to deductivism. Spinoza did not, like Descartes and many others, talk at great length about his method and the gains its adoption would bring. Rather, he simply used a particular method, and he used it realistically, comprehensively, and consistently. His workmanlike approach can be seen at the beginning of the *Ethics*, with his seven initial axioms closely resembling the simplicity and self-evidence of the Euclidean postulates.

The proofs, also, give an initial appearance of being genuine proofs and not, as they sometimes seem in Descartes, brief essays in support of contentious views. Thus, from the two axioms:

Axiom 4. The knowledge of an effect depends on and involves the knowledge of a cause.
Axiom 5. Things which have nothing in common cannot be understood, the one by means of the other,

Spinoza tries to deduce:

> Proposition 3. Things which have nothing in common cannot one be the cause of the other.

The proof itself is obvious. Assume two things which have nothing in common. Then, by Axiom 5, we cannot understand one in terms of the other. Consequently, by Axiom 4, neither can be the cause of the other. Therefore, we have proved Proposition 3.

But the proof depends upon the soundness of the axioms. However impeccable the rigour of the logic employed, if the axioms are at all doubtful, then the system itself will be suspect. In this Spinoza has fared no better than many another system-builder.[22] Is, for example, the already quoted Axiom 4 really acceptable? To have some knowledge of an *effect* do I really need to have knowledge of the cause? I know Newton had a breakdown in 1693, and I also know that Vesalius died in mysterious circumstances in 1564. The causes of these events are unknown to me and, I fear, anyone else. Many things are known about past catastrophes and present diseases without their causes having yet been identified.

Again, framers of deductive systems, though intent on formulating axioms of unquestionable truth, frequently find themselves repeating no more than the common presuppositions of their day. Thus Spinoza begins Part II of the *Ethics* with a further set of axioms of which Axiom 1 states: 'All bodies are either in motion or rest.'[23] The axiom, with its assumption that motion and rest are exclusive and exhaustive, is clearly pre-Newtonian. If the motion is uniform then the terms 'motion' and 'state of rest' do not exclude each other; if the motion referred to is not uniform then the distinction is not exhaustive. Reliance upon Axiom 1 allowed Spinoza to infer incorrectly as Lemma 3 that 'A body in motion or at rest must be determined to motion or rest by another body.'[24] It is in fact, as today's schoolchildren well know, *change* of motion or acceleration that needs to be determined in this manner.

Leibniz

A similar attraction to deductivism can be seen in Leibniz. As a student at Jena in 1663 he was much influenced by the mathematician Erhard Weigel (1625–99) who had sought to show that Aristotelianism, presented deductively, could be reconciled with modern philosophy. Leibniz was so impressed by Weigel's approach that he abandoned the traditional scholastic methods of his day in favour of mathematical demonstration. An early fruit of this approach was his *De arte combinatoria* (1666), and his curious *Specimen demonstrationum politicarum* (1669). In the latter work he demonstrated in sixty propositions which of four candidates should be elected to the Polish throne after the abdication of John II in 1668. Calculation, Leibniz revealed, indicated that the Palsgrave von Neuburg was the rightful choice, a candidate also supported by the Elector of Mainz, the patron of Leibniz. A later demonstration in 1671 established the conclusion, equally welcome to his patrons, that Louis XIV should remove his armies from the Low Countries to Egypt in order to fight the Turks. Louis, however, seems to have been unaware of the demands of logic and carried on serving French interests by preferring to pillage his neighbours.[25]

Leibniz had in fact, like many of his contemporaries, a far grander vision than the political manoeuvres described above. His aim was to develop a language, a calculus, or, in the jargon of the day, a Universal Characteristic, which would permit the routine solution of all problems. Just as numerical calculations could be made, accurately, effortlessly, and without disagreement, once quantities were expressed in the language of number, so too, it was expected, similar advances would be made in other areas once appropriate symbols had been found in which to express our thoughts. With appropriate rules, reasoning would become as mechanical and reliable as calculation.

Proposals of this kind were presented by Leibniz in his *Lingua generalis* (1678). Simple ideas were represented by prime numbers, and complex ideas were formed from them as their products. But, Leibniz soon realised, before a genuine calculus of thought could be developed, all simple ideas would have to be listed. Attention was therefore directed to the compilation of an exhaustive *Encyclopedia*. As with so many of his other schemes Leibniz rapidly abandoned the project for more pressing duties, and little remains of the proposed *Encyclopedia* other than the table of contents.

The ethical calculus

Several philosophers of the period set out to show that the propositions of morality could also be presented deductively. Weigel apparently led the way with a treatise entitled *Ethica Euclidea.* Locke too was interested in such a project. In his *Essay*, for example, he wrote of the proposition 'Where there is no property there is no injustice' that it was as certain as anything in Euclid. He went on to insist that morality as a whole was capable of demonstration, that: 'From self-evident propositions, by necessary consequences, as incontestable as those in mathematics, the measures of right and wrong might be made out, to any that will apply himself with the same indifference and attention to the one as he does to the other of the sciences.'[26] Locke did not in fact pursue his proposal. One who did, however, was Francis Hutcheson in his *Inquiry into the Original of our Ideas of Beauty and Virtue* (1725).[27]

Hutcheson did, literally, seek to introduce 'Mathematical Calculation' into morality. He recognised that before any computation of the morality of any actions could begin, it was necessary to establish a number of axioms. To convey something of the flavour of Hutcheson's approach his first axiom will be quoted in full:

> Axiom 1. The moral importance of any Agent, or the Quantity of publick Good produc'd by him, is in a compound Ratio of his Benevolence and Abilitys: or [by substituting initial letters of the words – Ability, Benevolence, Moment of Good] $M = B \times A$.

That is, the morality of an action consists in the product of the agent's benevolence and ability or, other things being equal, the more able and benevolent an agent, the better his behaviour.

Letting I stand for 'Moment of Private Good or Interest', and S for self-love, Hutcheson added a second axiom:

> Axiom 2. $I = S \times A$.

And from the axioms, using simple algebra, Hutcheson derived such theorems as:

$B = \frac{M}{A}$ (divide both sides of Axiom 1 by A).

In words, the benevolence of an agent varies directly as the good he produces and inversely as his ability. Something of the idiocy of this approach can be seen in the consequence that a well-meaning but hopelessly incompetent saint could be as benevolent as the efficient villain who did manage to carry out the few good deeds he attempted.

More generally, Hutcheson's ethical calculus shows one further demand of deductivism which cannot be safely ignored: the total dependence of a system upon its axiomatic foundations. If these contain any ambiguities, obscurities, or uncertainties, then similar features will rapidly appear in the derived theorems. Alternatively, if the axioms are too bland, they will permit no more than the deduction of equally bland conclusions.

Carnap and modern deductivism

The present century has seen an enormous growth in logical range and power. Many new techniques are now available to the logician and, at last, he has acquired a clear understanding of the nature of a formal system, and of what it is to prove something. And, just as the physicists of the seventeenth century required the more sophisticated mathematical techniques developed by Newton and Leibniz to provide an adequate analysis of motion, perhaps also the more powerful techniques available to the modern logician will permit equally important advances to be made in philosophy. Consequently, modern philosophers can feel as little constrained by the failures of their predecessors as Newton need have been by Aristotle's inability to develop a coherent theory of motion.

One of the first to put the new logic to the test was Rudolf Carnap. He had studied maths and philosophy at Jena under Frege and in 1913 he began a physics doctorate on the emission of electrons. But, as he later recorded, the project was technically difficult, and quite beyond his experimental skills. He turned instead to more theoretical issues and, on his return from the war, under the influence of Russell and Whitehead's *Principia Mathematica*, he began to work on the axiomatic foundations of kinematics. The work received little recognition. Carnap found himself ignored by the philosophers as a physicist and dismissed by the physicists as a philosopher. He moved to Vienna in 1926 at the invitation of Moritz Schlick, one of the founders of the school of logical positivism, and soon afterwards published his first major work, *Der Logische Aufbau der Welt* (1928).[28]

As a logical empiricist, Carnap regarded experience as fundamental. He undertook, therefore, to show how from the simplest elements of our experience, using only logical machinery, all other aspects of our thought and experience could be constructed. After all, Russell had done no less for the concepts of mathematics.

Carnap took as his single primitive term the notion of similarity recognition. Two colour shades, for example, or two smells, could be recognised as similar and consequently assigned to the same similarity class. At this point Carnap began to move up the constructive hierarchy. Similarity classes could be used to define the notion of a quality class, which in turn allowed him to define sense classes, and eventually, at least in theory, led to the construction of the class of physical objects. In this manner not only could such mental objects as thoughts and other minds be constructed, but cultural objects such as Calcutta, physical objects such as the Pacific Ocean, and such concepts of science as force, neurons, and magnetic fields.

The work, however, remained incomplete. Carnap had sketched a programme rather than carried out, in the manner of Russell and Whitehead, a detailed and rigorous derivation of the concepts of experience. He himself rapidly became dissatisfied with the work, and came to favour a more physicalist approach to the phenomenalism of the *Aufbau*. Although widely praised for its technical ingenuity, the work has been largely ignored. One of its few admirers, Nelson Goodman, has conceded that it is 'a crystallisation of much that is widely regarded as worst in 20th century philosophy'.

The weakness of deductivism

Some of the failures of deductivism have already been mentioned. While it has proved enormously creative in mathematics, no such rewards seem to have been collected by philosophers. Rather than being used to facilitate discovery, deductive techniques have been too often used to demonstrate the truth of views already established on other grounds. Just as Leibniz could be relied on to provide logical proofs of propositions acceptable to his political masters, philosophers have first decided what needs to be proved before constructing their systems. Logic in this context belongs more to rhetoric than to mathematics.

Again there is a notable tendency to talk a good deal about the virtues of deductivism while seldom actually presenting detailed deductive proofs. More often than not, as with Descartes, it is the language rather than the reality of deductivism which is pursued. Or alternatively, as with Carnap and Leibniz, the programme is started but never completed. The reason for this latter withdrawal is clear. While the initial stages of demonstrations tend to be simple and promising, deductive systems have a tendency to become messy, complex, and clogged up with ever more implausible assumptions. Eventually, sketches rather than proofs are offered, and amidst an increasing number of difficulties, the system is abandoned.

And when a Spinoza does attempt to create a genuine deductive system, a further difficulty arises. Such systems ultimately depend on the soundness of their initial assumptions. Unless quite innocuous, they will probably arouse considerable opposition; while if uncontentious, they will permit no more than the deduction of trivia.

The point links with a more general objection raised by S. E. Toulmin.[29] Concepts to the deductivist are timeless. While this may well hold for such notions as triangles and parallel lines, it fails completely when applied to the familiar concepts of cause, explanation, nature, and liberty. Frege has expressed the need to gain knowledge of a concept in its purest form, one from which all 'irrelevant accretions' have been stripped. Yet it is more than likely that many of those 'irrelevant accretions' are precisely the features which have endowed the concept with philosophical interest. Further, such a timeless approach ignores the important fact that concepts have histories; they change and develop, expand and contract. To identify one aspect of the concept as its purest feature is to run the risk Spinoza ran, and to find that supposed logical analyses are no more than the unquestioned assumptions of one particular age.

Newtonianism

Not all philosophers were impressed by deductivism. For them an alternative, and as compelling a model, became available with the unprecedented success of Newton's *Principia* (1687). Newton had managed, for the first time, to establish a basic cosmology able to deal with all phenomena. Celestial as well as terrestrial phenomena, both uniform and accelerated motion, the simple and the complex, the macroscopic and the microscopic, all were embraced by the comprehensive framework of the Newtonian cosmos.[30]

Principia begins with three laws of motion and eight definitions. All the laws and six of the definitions refer to forces of some kind. Much of the impact of the Newtonian revolution was gained by showing how forces could be used to explain the motions of the planets, the operations of the tides, the orbits of comets, the refraction of light, and, among many other phenomena, basic chemical reactions. The forces contemplated by Newton were not the vague and convenient operations favoured by generations of magi. Rather, they were controlled by specific quantitative laws, and checkable by experiments and observations. Thus the force operating between the earth and the moon, gravity, varied directly with their masses and inversely as the square of distance separating them. Whether the lunar orbit did behave in this manner was checked scrupulously by Newton and the results recorded.[31]

The impact of Newton's ideas can be seen most clearly in the manner in which scholars adopted his methods, concepts, and language. One of the earliest and strangest reworkings of Newton was the characteristically titled *Theologia christianae principia mathematica* (1699), written by the mathematician John Craig. Properly used, Craig argued, mathematical principles could illuminate Christian theology. Like Newton he began with three laws. Compare the first law of each:

Newton: Every body continues in its state of rest, or of uniform motion in a right line, unless it is compelled to change that state by forces impressed upon it.

Craig: Every man endeavours to prolong pleasure in his mind, to increase it, or to persevere in a state of pleasure.

From his laws and definitions Craig went on to derive such propositions as:

> Velocities of suspicion produced in equal periods of time increase in arithmetical progression.

Behind this apparent parody of Newton there lay a serious if misguided point. Craig started from Luke 18: 8: 'Nevertheless when the Son of Man cometh, shall he find faith on earth.' From this he inferred that when Christ returned there would be at least one believer on earth. Craig also held that belief in historical events diminished as they receded further into the past. If we could determine the initial degree of belief Christ's followers invested in him, and the rate at which such beliefs diminish, then, we would be able to establish an upper limit for the time of Christ's second coming. Eventually Craig established that as the probability of belief in Christ would decline to zero by 3150, he must return before then.

Hume and the associationists

Philosophers were also attracted to Newton's system, and none more so than David Hume. He had begun his *Treatise of Human Nature* (1738) by noting, like many before and since, 'the present imperfect condition of the sciences'. As it was evident to Hume that 'all the sciences have a relation . . . to the human mind', it was equally clear that only by developing 'the science of MAN' could success ever be gained in 'our philosophical researches'. Hume's *Treatise* sought to present such a 'science' of man.

He began in a Lockean manner by dividing the perceptions of the mind into impressions and ideas. Ideas were thought to be faint images of the impressions met with in thinking and perceiving. All simple ideas such as my memory of red were derived from a corresponding simple impression. There could, thus, be nothing in the mind which could not be ultimately traceable to an initial impression. So far Hume had done no more than identify the contents of the mind; it remained to establish a dynamics to explain its operations. At this point Hume seems to have turned to Newton for inspiration. 'Were ideas entirely loose', he noted, 'chance alone would join them; and it is impossible the same simple ideas should fall regularly into complex ones (as they commonly do), without some bond of union among them.'[32] What, it was natural to ask, was the force or forces which gave to ideas some 'union or cohesion'? Three principles were identified by Hume as responsible for the association of ideas: resemblance, contiguity, and cause and effect. Examples, absent from the *Treatise*, were offered in the *Enquiries*:

A picture naturally leads our thought to the original [resemblance]: the mention of one apartment in a building naturally introduces an enquiry or discourse concerning the others [contiguity]: and if we think of a wound, we can scarcely forbear reflecting on the pain which follows it [cause and effect].[33]

It may or may not have been a coincidence that Hume, like Newton, based his work on three principles; of the source of Hume's approach, however, there can be no doubt. Having introduced the principles behind the association of ideas, Hume went on to comment: 'Here is a kind of attraction, which in the mental world will be found to have as extraordinary effects as in the natural, and to show itself in as many and as various forms.'[34] To an eighteenth-century reader 'attraction' could refer to just one subject, the physics of *Principia*. The point was repeated in the *Enquiries*. In the first section he compared philosophy to astronomy. Scientists had long been content with describing 'the true motions, order, and magnitude of the heavenly bodies'. But, at last, a philosopher arose who succeeded in determining 'the laws and forces, by which the planets are governed and directed'. In the same manner, Hume suggested, 'equal success' could be attained 'concerning the mental powers and economy'.

Association, Hume stressed, did not deal in 'inseparable connections'; it was, rather, 'a gentle force which commonly prevails'. Also, like a good Newtonian, he would restrain 'the intemperate desire of searching into causes', and he would shun 'obscure and uncertain speculation'. Consequently, he would restrict himself to 'examining the effects' rather than 'the causes of his principle'.

Even more explicit in his dependence on Newton was David Hartley (1705–57). His work, he noted in his *Observation on Man* (1749), was based on hints 'concerning the performance of sensation and motion' taken from Newton's *Principia* and *Opticks*. In the General Scholium Newton added to the second edition of *Principia* (1713) he argued:

all sensation is excited, and the members of animal bodies move at the command of the will, namely, by the vibrations of this spirit, mutually propagated along the solid filaments of the nerves, from the outwards organs of sense to the brain, and from the brain to the muscle.

Hartley sought to forge Newton's vibrations and Hume's principles of association into a coherent whole. '*Vibrations* should infer *associations* as their effect', he judged, 'and *association* point to *vibrations* as its cause.'[35]

A century later associationists could still be found formulating their various laws. John Stuart Mill, for example, in his *A System of Logic* (1843), accepted that the laws of the mind were associationist. Three laws were identified: similarity, contiguity, and intensity.[36] Later in the century Alexander Bain, the founder of the journal *Mind*, could be found formulating in his *Mental and Moral Science* (1872) the three laws of contiguous adhesion, similarity, and compound association.[37]

Reid and the philosophy of common sense

Beginning in Aberdeen and Glasgow in the late eighteenth century there arose, under the initial guidance of Thomas Reid (1710–96), the so-called Scottish school of common sense philosophy.[38] Among other views they were united by a deep suspicion of traditional empiricism as it had been developed by Locke, Berkeley, and Hume. Part of the original inspiration for this suspicion arose from their study of Newton, particularly his methodological writings.

The classic statement of Newton's approach to science is to be found in the four 'Rules of Reasoning in Philosophy' which introduce Book III of *Principia*. According to the first rule 'We are to admit no more causes of natural things than such as are both true and sufficient to explain their appearances.' Taken in conjunction with Newton's well-known insistence that he 'feigned no hypotheses', the rule was taken by Reid as an unmistakable ban on all forms of hypothetical reasoning. Consequently, presented with any philosophical claim that we have discovered the cause of any effect relating either to matter or mind, we must, Reid insisted, 'first consider whether there is sufficient evidence that the cause he assigns does really exist. If there is not, reject it with disdain, as a fiction which ought to have no place in genuine philosophy.' Once satisfied that the cause exists, Reid conceded, we can then and only then go on to ask whether the cause did operate as described. 'Unless it has these two conditions', Reid concluded, then it will be 'good for nothing'. Hypotheses, he repeated, should be treated with contempt 'in every branch of philosophy'.

Reid thus shows the reverse side of the Newtonian influence on philosophy. Whereas Newton stimulated Hume to develop new theories, Reid was led to argue that most theories must be discarded; whereas Hume saw ways to liberate philosophy under Newton's influence, Reid saw only ways in which it could be confined. The dangers of allowing the methods of science to control philosophy, or any other discipline for that matter, are all too apparent in Reid's narrow interpretation of Newton.

As read by Reid, only *true* hypotheses are allowed in philosophy, as in science. Applied to science itself, it would immediately destroy the bulk of scientific theory from Reid's day onwards. Dalton, for example, was in no position to show that his atoms really did exist when he first published his views in 1808. If he had followed Reid's advice he would not even have formulated his atomic hypothesis.

It should be remembered that Reid was writing at a time when science had begun to embrace a growing number of, in the language of the day, imponderables. Bodies grew hot when they absorbed caloric, they were combustible because they were rich in phlogiston, and attracted iron because they contained an abundance of magnetic effluvia. Asked to exhibit and measure these new bodies, scientists explained away their failure by arguing that substances like caloric and phlogiston were too rarefied, too imponderable, to show up in the crude scales of the chemist. The ploy could be used uncritically, as Reid was well aware. The point remains that it would often be preferable to accept unproved hypotheses and see just how fruitful they were.

Attempts to apply something like Reid's criteria by a more recent generation of positivists has led to similar difficulties. Logical positivists were once ready to argue that the existence of entities should only be accepted when the claim could be verified in sense experience. In this manner we can establish that chairs, but not unicorns, really do exist. But what, though, of the claims for the existence of neutrons, magnetic monopoles, and the unconscious? Where is the evidence for their existence? Like Dalton's atoms, evidence for the existence of such objects was to be found more at the theoretical level than in sense experience.

Alternative views: Hartley and Le Sage

Newton's influence on philosophy could be enormously variable. Not all eighteenth-century philosophers were attracted to the sparse universe favoured by Reid and his colleagues. One, for example, who objected strenuously was David Hartley, physician, psychologist, and author of *Observations on Man* (1749).[39]

Hartley began from 'hints' on sensation and motion he claimed to have found in the General Scholium Newton added to the second edition of *Principia* (1713), and to the 'Queries' added to the various editions of the *Opticks*. At several places Newton spoke of vibrations propagated along the optic nerve into the brain. He went on to ask:

Do not the most refrangible rays excite the shortest Vibrations for making a sensation of deep violet, the least refrangible the largest for making a Sensation of deep red, and the several intermediate sorts of Rays, Vibrations of several intermediate bignesses to make Sensations of several intermediate Colours?[40]

Hearing, too, Newton suggested, could result from vibrations 'excited in the auditory nerves'. Animal motion, he also proposed, could arise from vibrations excited by the will in the brain and 'propagated thence through the solid pellucid, and uniform Capillamenta of the Nerves into the Muscles, for contracting and dilating them'.[41]

Newton's proposals were relatively modest and tiresomely tentative. Hartley was less restrained. In essence he sought to link Newtonian nervous vibrations to the associationism of Locke and Hume and account for virtually all mental phenomena, and much of our basic physiology as well. 'Vibrations should infer associations as their effect, associations point to vibrations as its cause', he insisted.[42] Indeed, little is missing from Hartley's programme. Vibrations of an appropriate kind are offered to account for pleasure, pain, voluntary and involuntary movements, sexual desire, the sensations of heat, taste, smell, sight, hearing, dreams, and much more besides.

Yet, derived from Newtonian sources though they may have been, Hartley was acutely aware that such a free use of nervous vibrations was incompatible with the received canons of Newtonian methodology. He responded frankly and somewhat pragmatically. It was enough, he argued, that his hypotheses really did succeed in explaining a wide variety of phenomena. In defence of his view Hartley used the familiar notion of a code. Presented with a passage in code, the cryptographer begins by making a number of assumptions. He will then try, with his assumed key, to make sense of the coded text. If he succeeds in establishing in this manner an intelligible and appropriate plain text, he will conclude that his original assumptions were true.

The weakness of Hartley's position is only too apparent. The analogy with cryptography fails to work because for coded passages of any length there is likely to be only one solution. It is therefore sufficient in this case to find a key. It immediately follows that it must be the right key. The less constrained phenomena of nature are not quite so hospitable. Hartley's claim, therefore, that he framed hypotheses which, whatever their faults, did succeed in explaining a wide variety of phenomena, was likely to be greeted by similar claims from his numerous rivals. As writers of detective fiction continue to demonstrate, and as the appeal courts of Britain regularly allow, there are invariably several hypotheses capable of explaining all the facts. The fundamental question for scientists and philosophers remains, as for judges, which of the competing hypotheses is true.

Another who resorted to the method of hypothesis was the Swiss scientist George Le Sage. He attempted to do in public what Newton had only dared to do in private, namely to construct a mechanical model of gravity. He supposed there to be an ethereal medium surrounding all bodies. Also, he allowed it to consist of solid particles moving randomly in all directions. A single body will, therefore, receive impacts from all directions and consequently remain motionless. If, however, the universe contains two bodies, A and B, such as the moon and the earth, each will partially shield the other. More impacts will, therefore, be received on one side and the bodies will tend to move to each other. What appears to be a rather mysterious case of gravitational attraction between the earth and the moon is really, Le Sage argued, merely the result of pushes they receive from otherwise imperceptible particles.

The reaction to Le Sage's work was invariably hostile. Further, the hostility, as Laudan has noted, was based on philosophical rather than physical grounds.[43] The mathematical physicist Roger Boscovich argued that Le Sage's hypothesis was purely 'arbitrary'. The mathematician Euler informed Le Sage that he felt 'une très grande répugnance pour vos corpuscles'. In the face of such opposition Le Sage abandoned the full account of his system that he was preparing for the press. In its place he chose to defend his use of hypotheses and 'agens imperceptibles'. His main argument, that hypotheses based on a wide and varied sample deserved serious consideration, found little support.

Evolution

In more recent times both Popper and Toulmin have presented accounts of knowledge based on evolutionary models.[44] The approach is far from new. T. H. Huxley, for example, in an address in 1880 on the twenty-first anniversary of Darwin's *Origin of Species* argued that 'The struggle for existence holds as much in the intellectual as in the physical world. A theory is a species of thinking, and its right to exist is coextensive with its power of resisting extinction by its rivals.'[45] Similar views have been expressed by such divergent thinkers as Herbert Spencer, Ernst Mach and Teilhard de Chardin.

In its contemporary form, however, it surfaced in Popper's 1961 Spencer lecture, 'Evolution and the Tree of Knowledge'. The growth of our knowledge, Popper claimed:

is the result of a process closely resembling what Darwin called 'natural selection'; that is, *the natural selection of hypotheses*: our knowledge consists, at every moment, of those hypotheses which have shown their (comparative) fitness by surviving so far in their struggle for existence; a competitive struggle which eliminates those hypotheses which are unfit.[46]

Such talk, Popper emphasised, was not meant to be taken metaphorically. But is talk of this kind at all plausible? And, further, is anything to be gained by its adoption?

It is difficult to conceive of a more wildly inappropriate manner in which to talk of the survival of theories and hypotheses. Organs which have survived have clearly shown their comparative fitness. Survival and fitness are simply two ways to make the same point. Organs do not develop which make animals less fit to survive; and those who do survive do so because, in comparative terms, they are fitter to survive. Theories differ. They may survive for social reasons, through propaganda, censorship, and other political tools. Alternatively, they may die out not because they are unfit, but through stupidity, prejudice, and corruption. It is legitimate to say of a hypothesis that, though it has survived a long time, it should have been discarded long ago; equally, we can say of a theory which failed to survive, that it should never have been rejected. Less, if any, sense is however made of the claims that the trilobites should have survived, and that lizards should have long been extinct. The truth is that *fit* theories can die and *unfit* ones survive. The same cannot be said of organisms within the confines of Darwinism.

Further, theories, unlike organisms, can be revived. Species which become extinct are gone forever. A theory dismissed, the heliocentrism proposed by Aristarchus and others in antiquity, can be found under consideration 1500 years later by Copernicus. While we can say of a species that it was unfit to survive, more prudence is needed for theories. Theories, unlike species, can be dormant.

There is also the issue of novelty. Nothing, of course, in the evolutionary record is entirely novel. When palaeontologists speak of novel forms bursting through, they have in mind species lasting tens rather than hundreds of millions of years. Species have ancestries; they do not emerge from nothing. Theories can sometimes be completely original, unanticipated and without ancestors.

Finally, what of competition? Species compete for the same resources. Plankton can be eaten only once. What of theories? Did the wave and corpuscular theories really compete? Competition is competition for the same resources. The wolf does not compete with the finch, nor do the wolves of Canada compete with the wolves of Siberia. Popper's assumption is that just as competition between species leads to improvement, to the development of fitter species, so too does competition between theories. In the conflict weaker theories will be discarded and fitter theories will conquer. But history shows no such guarantee. In the competition between theories the weaker one has won as often as the stronger. Aristarchus lost to Ptolemy for 1500 years. Galileo lost to the church, and generations of past anatomists have argued that black men and all women of any society are the intellectual inferiors of white men.

Alternatively, competition may fail to resolve the issue, and rival theories may coexist for centuries. This is quite unlike nature, where the 'one niche: one species' rule holds. Keep paramecia in a test tube and, given sufficient nutrients, the population will flourish. Put in two species, however, and the outcome will always be the same, namely the complete extermination of one of the species.[47] Only if species can find a niche of their own can they hope to survive. Yet today we see belief in astrology is held as widely and as firmly as belief in astronomy. Indeed, they have happily coexisted for well over two millennia.

Selection and the M-16

Belief in theories, it could be argued, is always likely to be untypical of beliefs in general. Human stubbornness and perverseness will always guarantee some support for the black-is-white party. What though of the real world, where choice could be a matter of survival? Surely, here, competition will ensure the selection of whatever is most effective. Especially so if the choice is between two rifles and is being made by an advanced, technical army in time of war.

Such a choice faced the US army in the 1960s as it considered how to equip its forces in Vietnam. Quite amazingly, the Pentagon rejected the AR-15, 'the most reliable, and the most lethal, infantry rifle ever invented', in favour of the M-16, a rifle which repeatedly misfired and jammed. How this strange choice was made became the subject of a Congressional committee enquiry.[48]

The basic difficulty was the fact that the US army could not restrain itself from imposing on the already deadly AR-15 their own standards. These turned it into the hazardous M-16. They insisted that a manual bolt closure should be added, the barrel twist increased from 1 in 14 to 1 in 12, and that the Armalite IMR powder be replaced with the army's own ball powder. Behind the changes there lay two dangerous assumptions. The first was that no private company like Armalite could be expected to be completely familiar with military needs. Therefore whatever they offered would have to be adapted. Secondly, it was assumed that the rifle would be aimed and fired at distant targets. This meant it would have to be accurate at 600 yards, and suitable for a marksman to use. This led to the numerous changes demanded by the Pentagon.

But it turned out that a quite different rifle was needed in the forests of South East Asia, one which fired many unaimed shots, reliably, and often at a short distance. Instead, the troops received a rifle which in fifty rounds of fire could double-feed and jam some fourteen times.

This, one is tempted to say, is how both theories and objects not uncommonly survive in the intellectual and social world. In this world survival is the result of a number of factors, few of which will have anything to do with fitness and effectiveness. Further, if the survival of rifles is determined in this manner, it is unlikely that abstract theories will be selected on any better grounds.

Two objections to this view are frequently met with. It is first claimed that survival alone *shows* fitness. The M-16 survived because it appealed more to the generals than the AR-15. The trait of being attractive to the powerful is just as much a survival trait as firepower. The difficulty of this view is that it makes the claim trivially true. Whatever survives *is* fittest! The converse, in fact, of Darwinism. In this manner Pekinese survive, not because they are fit enough to cope with life, but because they are cute enough to appeal to a class of lonely women. It is sufficient to point out that we have come a long way from the sense of fitness Popper attributed to successful theories.

Secondly, it is pointed out, evolution is a continuous process. The faults of the M-16, apparent to Congress, would no doubt have been corrected in due course. Faults can always be found in artifacts and organisms alike; the issue is whether or not they can be eliminated. What survives is the final model, not the prototype. Eohippus made room for later and better forms just as naturally as Newton improved on the earlier theories of his youth. The argument fails because it forgets that evolution can only be about the past and the present. It is true that artifacts and organisms can improve; that the fit can become fitter. They can also, however, just as easily stagnate and, even, deteriorate. To assume otherwise is simply to beg the very question at issue.

PART IV

SOME CRITICAL STYLES

8

IS SCIENCE OPEN TO PHILOSOPHICAL CRITICISM?

> But he proposed himself to be a Philosopher and not an Astronomer, and therefore I have no more to say to him but that Astronomye is ill served when it must be ordered by whymsies of Philosophy.
>
> John Flamsteed, the first Astronomer-Royal,
> unpublished letter about Edmond Halley (1678)

Critical styles

There has long been in the West a tradition critical of the social and spiritual consequences of science. As early as the fourth century BC Epicurus advised that it would be far better to follow the myths of the philosophers than to be a slave to the fatalism propounded by the physicists. Many since, such as St Augustine, have emphasised the essential inadequacy of science. In his *Confessions* he vividly recalled how he had been led from Manichaeism, an early dualist heresy, by the impressive power of astronomers to explain and predict celestial phenomena. Yet he soon realised it was insufficient to be able 'to number stars and grains of sand', or 'to trace the paths of planets'; only knowledge of God, he insisted, could ever bring real peace and happiness.[1] The point has been repeated endlessly. Matthew Arnold, for example, in almost identical words, in his 1882 Rede lecture on 'Literature and Science' saw science as essentially 'unsatisfying' and 'wearying'. While science could provide much accurate information about plants, animals, stones, and stars, he complained, it failed to relate such knowledge to 'our sense of conduct, our sense of beauty'.[2]

Others have found science to be not just inadequate but positively harmful. Keats voiced the complaint that science has destroyed such magical visions as the rainbow by reducing them to just one more entry in the already overcrowded 'dull catalogue of common things'. At the same time William Blake was uttering prophetic warnings about the disastrous effects the 'Looms of Locke' and the 'Water-wheels of Newton' would have upon the lives of men. Equally familiar has been the complaint of Kierkegaard against the scientific 'blasphemy' of seeking to describe man in the mechanistic language used to talk of plants, animals, and the heavens. It is echoed in the charge of Theodore Roszak that the much valued objectivity of science can destroy human values by treating people as things.[3]

There can also be found in western thought a philosophical tradition critical of the *intellectual* claims of science. At first sight the notion of philosophers offering themselves as serious critics of science is likely to appear both absurd and pretentious. And, indeed, the folklore of philosophy is rich with tales of foolhardy and arrogant philosophical critics who have, in Popper's phrase, 'drawn real physical rabbits out of purely metaphysical silk-hats'.[4] One of the most notable such conjurors was Cesare Cremonini, Aristotelian philosopher and colleague of Galileo at Padua. In popular tradition Cremonini is presented as dogmatically insisting on the superiority of the Aristotelian description of the heavens over reports of novel phenomena observed by Galileo through the newly invented telescope. Offered the chance to see for himself, Cremonini supposedly declined, preferring the authority of Aristotle to the evidence of his own senses. Consequently he rejected as fictitious Galileo's reported observations of the satellites of Jupiter and the mountains of the moon.[5]

Equal scorn has been directed against Hegel for the claim in his *De Orbitis Planetarum* (1801) that no planet could be located between Mars and Jupiter. Even before his text appeared, the minor planet Ceres had been observed by Giuseppe Piazzi between these very planets. Franz Zach, Director of the Gotha Observatory, accused Hegel of 'literary vandalism' and suggested that he and his colleagues should 'first learn before they teach'.[6]

Nor, apparently, are twentieth-century philosophers any less arrogant and misguided than their predecessors. One such encounter between scientists and philosophers was witnessed by the young Roy Harrod and described by him in his memoir on the physicist Lord Cherwell.[7] Some time in the 1920s Cherwell addressed a meeting of Oxford philosophers on the work of Einstein. The first to reply, Harrod recorded, was J. A. Smith, Waynflete Professor of Metaphysics. Relativity, he insisted, was inconsistent. Several physicists in the audience quickly told him that he had made a number of unnecessary assumptions. Looking 'extremely vexed', Smith 'knocked out his pipe abruptly in the fireplace and sat down'.

He was followed by H. W. Joseph, a sharper and more persistent philosopher. To the discernible dismay of his audience he began to read from 'a thick wad of manuscript'. Space was Euclidean, he argued, and consequently Einstein's views could not possibly be correct. And further, he asked, what could space be curved in? Yet, however cogently Joseph appeared to argue, it remained clear to Harrod that he:

knew nothing whatever about the Theory of Relativity. Since it was so evident to him that the conclusion reached was untrue, on quite different and sufficient philosophical grounds, there was no need for him to bother with the reasons why certain persons had been induced to frame such a theory.

Cherwell was unable to persuade the bulk of the audience in Einstein's favour; as they left, Harrod tells us, 'The Wykehamist Greats men were jubilant; a scientific professor had been torn to pieces; the Theory of Relativity had been shown to be untrue.'

Not all philosophical surveys of science are as ill-founded as the polemics of Cremonini, Hegel, and Joseph. The role of philosophical critics of science is a complex one and at least three quite distinct critical styles can be distinguished. The boldest and rarest approach concerns itself with the very viability of science itself. The methods, theories, results, and even the spirit of science are challenged by such radical critics. Most famous contemporary representative of this genre is the self-proclaimed anarchist philosopher Paul Feyerabend. He has made the apparently serious claims that astrology, voodoo, witchcraft, rain-dancing, and much more besides, are no less reasonable procedures than astronomy, medicine, and meteorology. When asked why he prefers to fly in aeroplanes rather than on broomsticks, he answers that, while he knows how to use aeroplanes, he is as a matter of fact unfamiliar with broomsticks and insufficiently interested to learn more about them. Feyerabend's controversial thesis that 'anything goes' and his critique of science will be considered later in the chapter.

A second approach accepts the autonomy and authority of science yet finds much to question in the application of certain well established scientific theories. In this case it is less science and more the writings of scientists that are commonly challenged. Perhaps the most famous example of this approach has been Susan Stebbing's formidable survey in her *Philosophy and the Physicists* (1937) of the popular writings of James Jeans and Arthur Eddington. Distinguished though they may have been as mathematical physicists, such of their works as *The Nature of the Physical World* (1928) and *The Mysterious Universe* (1930), Stebbing argued at length, were at best misleading and, all too often, extravagant nonsense. Like many scientists before and since, Eddington and Jeans were keen to show that nature was much stranger than their readers could have suspected. However strange and mysterious nature may be, Stebbing had little difficulty in showing that it lay elsewhere than in the imagination of Jeans and Eddington. Critical work of the Stebbing type will form the subject-matter of the next chapter.

As an exponent of a third and more straightforward critical style we can identify the philosopher who finds on purely philosophical grounds questionable views in the science of his day. It is certainly a persistent tradition in philosophy. In antiquity Aristotle devoted much of his *Metaphysics* to a survey of the physics of his predecessors, as did the second century AD sceptical philosopher Sextus Empiricus in his *Outlines of Pyrrhonism*. Today's philosophers may be less comprehensive; they are seldom less critical. Popper, for example, has written extensively over the years on a number of difficulties in quantum theory. He relates in his intellectual autobiography how, as a young philosopher, and as a realist, he had long felt unhappy with the interpretation of quantum theory favoured by Niels Bohr. They discussed their differences in Copenhagen in 1936. Popper concluded: 'I suspected that Bohr's theory was based on a very narrow view of what *understanding* could achieve. Bohr ... thought of understanding in terms of pictures and models . . . I developed an entirely different view.'[8] In 1982 Popper offered his most complete analysis of quantum physics in his *Quantum Theory and the Schism in Physics*.

From an earlier period we can mention Leibniz and his opposition to the vacuum. Although considerable evidence had been gathered by the early eighteenth century for the existence of the vacuum, Leibniz had no hesitation in rejecting it on purely philosophical grounds. In his correspondence with Samuel Clarke he offered two main arguments. In the first place, he insisted, God's perfection required a plenum, for 'the more matter there is, the more God has occasion to exercise his wisdom and power.'[9] And, secondly, ''Tis impossible there should be any principle to determine what proportion of matter there ought to be, out of all the possible degrees from a plenum to a vacuum.'[10] And in the absence of any such principle, then, by Leibniz's own principle of sufficient reason – 'nothing happens without a reason why it should be so, rather than otherwise' – the vacuum becomes an impossibility.

Many other examples could be listed. Rather than selecting from them in an arbitrary fashion a number of unrelated case histories, we will examine two issues in some detail. To begin with, the manner in which one exceptional philosopher, George Berkeley, reacted to many of the scientific issues of the day will be considered. And in contrast to this, later in the chapter, the treatment by many philosophers throughout the ages of the single issue of atomism will be examined with an eye to identifying the roles science and philosophy have played in its development. It is, however, with Berkeley that we will begin.

Berkeley

Today Berkeley is best known for his claim that *esse est percipi* – to be is to be perceived – and consequently as the philosopher who absurdly maintained that the tree in the quad would cease to be if no one was present to observe it. For many readers the robust good sense of Dr Johnson was sufficient to expose the errors of Berkeley's position. Boswell related how one day they were talking of Berkeley's 'ingenious sophistry to prove the non-existence of matter, and that everything is ideal'. Somewhat rashly Boswell commented that though Berkeley's views were clearly false, they were hard to refute. But, he continued, 'I never shall forget the alacrity with which Johnson answered, striking his foot with mighty force against a large stone, till he rebounded from it, "I refute it *thus*."'[11]

But to many of his contemporaries Berkeley was better known for the formidable attack he directed against the Newtonians, an attack based squarely on his own philosophical position. His critical talents were initially seen in his first major work, *An Essay Towards a New Theory of Vision* (1709).[12] The subject was a topical one, and one on which scientists had begun to feel, and with some confidence, that here was an important topic they could illuminate with their own specialised techniques. Eyes contained lenses, and the laws governing the passage of light through a lens were scientific laws. Therefore, it was felt, some of the more puzzling features of the visual process would be best answered in purely scientific terms.

One pressing problem, known since the time of Leonardo da Vinci, was concerned with the inverted retinal image. How did we ever manage to see an upright image, given that the image on our retina was inverted? Throughout the seventeenth century the techniques of geometrical optics were repeatedly applied to this and a number of similar problems. Kepler's *Ad Vitellionem paralipomena* (1604, *Additions to Witelo*), Descartes's *La Dioptrique* (1637), Barrow's *Lectiones* (1669–70), Molyneux's *Dioptrica nova* (1692), and even Newton's *Opticks* (1704) all sought to geometrise vision. And it was this tradition the young Berkeley dared to challenge.[13]

He began by considering the question of how we judge distance. One answer quoted by him as popular with 'the optic writers' emphasised the degree of divergence of the rays of light issuing from the object as they enter the eye, concluding: 'that point being judged nearest, which is seen by most diverging rays; and that remoter, which is seen by less diverging rays'.[14] Alternatively, it was proposed, we could judge the size of the angle made by the optic axes as they meet at the object viewed, with nearer objects making greater angles, and more distant objects smaller angles.

Berkeley happily accepted the principles of optics deployed. He was equally content to allow that there was 'a necessary connexion' between distance and measurement. None the less he was insistent, and rightly so, that such principles could never explain how we could judge an object's distance. The basic error was to suppose that 'men judge distance as they do of a conclusion in mathematics.'[15] But the lines and angles used by the mathematicians in their demonstrations were fictitious and therefore unperceivable. Were brutes and children, Berkeley asked, supposed to judge the distance of an object 'by virtue of geometry and demonstration'?

A similar objection was raised against the Cartesian account of inversion of the retinal image. It was held that the mind, 'perceiving an impulse of a ray of light on the upper part of the eye, considers this ray as coming in a direct line from the lower part of the object'. Asked why we should make such a curious judgement, the Cartesian tended to fall back on a favourite analogy. The situation was like:

A blind man, who holding in his hands two sticks that cross each other, doth with them touch the extremities of an object . . . It is certain, this man will judge that to be the upper part of the object, which he touches with the stick held in the undermost hand, and that to be the lower part of the object, which he touches with the stick in his uppermost hand.[16]

The analogy was, according to Molyneux, 'allowed by all men as satisfactory'. Not however by Berkeley. For, he argued, 'Crossing and tracing of the rays, is never thought on by children, idiots, or in truth by any other, save only those who have applied themselves to the study of optics.'[17]

Behind the obvious common sense of Berkeley's reply there lay a more theoretical worry. The only possible objects of knowledge for Berkeley were ideas. And ideas were things which could not exist unperceived. When something was not perceived immediately, such as another's pain, it could still be known indirectly through such other ideas as a distressed appearance. But if these secondary ideas were unperceived then they could never be a means of perceiving the primary idea. As it was conceded that the lines and angles described by 'optic writers' did not exist, they could neither be perceived nor form the basis for judgements of distance.

In this case Berkeley had accepted the facts put forward by the physicists and had merely challenged their interpretation of those facts. In a later work, *The Principles of Human Knowledge* (1710), he took the matter a stage further and began actually to challenge the facts and basic assumptions of the science of his day. His quarry was none other than Newton.

Throughout much of his *Principles* Berkeley had sought to demonstrate that the objects of human knowledge are ideas, and that ideas cannot exist unperceived. Consequently, the objects of human knowledge – be they houses, rivers, or mountains – cannot exist unperceived. Once having convinced himself of the validity of his basic argument he proceeded to apply it rigorously to all areas of experience. Accordingly, Berkeley began to consider 'natural philosophy and mathematics'.[18]

In the preface to *Principia* (1687) Newton had emphasised that 'the whole burden of natural philosophy seems to consist in this': 'from the phenomena of motions to investigate the forces of nature, and then from these forces to demonstrate the other phenomena'.[19] One force, that of gravitational attraction, had been identified, and with its aid Newton proceeded 'to deduce the motions of the planets, the comets, the moon, and the sea'. He suspected also that other forces as yet unidentified existed in nature and were responsible for such phenomena as chemical reactions, and the cohesion of solid bodies.

Newton's views on this matter conflicted quite sharply with those of Berkeley. The only efficient cause Berkeley would recognise was 'the will of a spirit'. He had argued at length that matter, being entirely inert, could be the cause of nothing. Equally inert and causally irrelevant were the 'mechanical causes', the primary qualities of Locke, such features as 'the figure, motion, weight' of bodies. Thus, for Berkeley, a noise heard was not 'the effect of this or that motion or collision of the ambient bodies, but the sign thereof'.

Having dismissed matter, he turned to Newton's attractive forces. He agreed that stones do fall to the earth, that bodies cohere, and that the sea swells to the moon. Yet, he asked, what enlightenment comes with 'being told this is done by attraction?' Nothing more surely than 'the *effect* itself'. It could just as easily be termed '*impulse*, or *protusion*, as *attraction*'. Before Berkeley would concede that anything is signified by the notion of attraction he would need to know 'the manner of the action whereby it is produced'. How, in fact, does the moon attract the seas? And what cause produces this attraction?[20]

Newton had in fact long been aware of this precise problem and in the General Scholium he added to the second edition of *Principia* (1713), he replied to the objection. He had not, he conceded, 'been able to discover the cause of those properties of gravity from phenomena, and I frame no hypotheses . . . it is enough that gravity does really exist, and acts according to the laws which we have explained.'[21] Nor was he any more successful in the several attempts he made to work out, in Berkeley's words, 'the *manner* of the action whereby it is produced'.

Berkeley followed with an equally damaging attack on Newton's reliance on the concepts of absolute space, time, and motion. Newton had said that absolute space 'in its own nature, without relation to anything external, remains always similar and immovable'. Such a concept was dismissed by Berkeley as absurd on the grounds that we could never have any idea of such a scheme. It would be impossible, he insisted, for us to 'even frame an idea of pure space exclusive of all body'. To say something exists is to say, for Berkeley, that it is perceived. How, then, can space be perceived? By linking it to the notion of resistance. The more resistance is felt by another body (the perceiver) then the less empty space there is taken to be. But, he concluded, if all bodies were excluded, including my own, then 'there could be no motion, and consequently no space.'[22]

Theological objections were also raised by Berkeley. If there was such a thing as 'pure space', would it not have to be God? Or, if not God, would we not have to admit that something other than God was 'eternal, uncreated, infinite, indivisible, immutable'?

A final attack was directed against the calculus of Newton and Leibniz and was sufficiently important to warrant its own monograph: *The Analyst . . . Addressed to an Infidel Mathematician* (1734). Not surprisingly, a philosopher who found tables and chairs hard to accept would be most unlikely to welcome some of the strange entities introduced by the founders of the calculus. Thus, Newton had spoken of a line or a curve as generated by the continuous motion of a point. The line, a flowing quantity, was referred to by Newton as a *fluent*, and the rate at which it flowed, the point's velocity, was described as the line's *fluxion*. Fluents could, and did, grow by 'indefinitely' small parts in 'indefinitely' small periods of time – *moments* in Newton's terminology. Within this approach there was much talk of *moments* being 'indefinitely little' and, consequently, they and any terms multiplied by them could be eliminated from the equations.[23]

Such notions appalled Berkeley. What were these *moments*? Were they something infinitesimally small? Or were they nothing? Berkeley posed his dilemma. If *moments* really were something, however small, they could not be ignored; if, however, they were nothing, then their presence could add nothing to the equations. And what of fluxions, Berkeley asked, 'the Velocities of evanescent Increments': 'They are neither finite Quantities, nor Quantities infinitely small, nor yet nothing. May we not call them the Ghosts of departed Quantities.'[24]

Berkeley's comments were taken seriously by the English Newtonians. Replies came quickly from James Jurin and Benjamin Robins. The most considered and authoritative response, however, was to be found in Colin Maclaurin's *Treatise on Fluxions* (1742), a work which attempted to derive the rules of the calculus in a rigorously geometrical manner. The infinitely small was abandoned as too bold 'a Postulatum for such a science as Geometry', and the fluxional notation appeared only towards the end of the work.

Berkeley's suspicions about the calculus proved to be well founded. Little sense could be made of infinitesimals as long as they were taken as ultimate quantities. A more rewarding strategy, seen early by Robins, was to define ultimate magnitudes as limits which they can approach to within any required degree of proximity. This approach, however, required the more rigorous skills of the nineteenth-century mathematician before the idea could be developed satisfactorily. Before then, unsurprisingly, mathematicians continued to operate with their workable, but suspect, techniques.

Two general points remain to be made. First, Berkeley was not merely an interested observer of the science of his day. We are all free to read accounts of new scientific theories and comment on them as we wish. Our views, however, can only be judged by the prevailing scientific standards of the day. And in these terms our views may be held to be of interest, or they may not. Berkeley's comments were made from outside science and appealed to a different set of values. He objected to Newtonian science not because he considered it to be bad science, but because he was convinced that it made no philosophical sense. He was no amateur scientist claiming to be more knowledgeable than the professionals. Rather, he was claiming that no amount of scientific evidence or expertise could ever convert what was philosophically inept into scientific truth.

Secondly, Berkeley's bold comments were taken seriously by his scientific contemporaries. Sufficiently seriously, that is, to meet with reasoned defence in place of the ridicule or indifference with which scientists more often respond to the criticism of outsiders. Further, scientists have continued to feel the need to cope with Berkeley's critique and sometimes, even, to favour his approach. Popper has spoken of him as a 'precursor of Mach and Einstein', while quantum theorists such as John Wheeler have begun to propose the very Berkeleyan-sounding principle, the so-called 'Participatory Anthropic Principle', that observers are necessary to bring the universe into being.[25] This will be considered in chapter 9.

Feyerabend

The most sustained and extreme reappraisal of the pretensions of science has come from Paul Feyerabend in his *Against Method* (1975) and *Science in a Free Society* (1978).[26] He begins with the apparently simple question: what makes modern science preferable to, say, Aristotelian metaphysics or Hopi cosmology? The most obvious response, that it was scientists and not metaphysicians or Hopi priests who put a man on the moon, fails to impress Feyerabend. For the argument to be convincing, he has insisted, two further premisses must be demonstrated. It must first be shown that 'no other view has ever produced anything comparable', and secondly that the achievements of science are genuine products of the theories and methods of science and not merely the accidental outcome of 'non-scientific agencies';[27] that, in short, the theories and methods of science are both effective and uniquely so. The points will be considered separately.

The effectiveness of science

Feyerabend begins his argument by asking whether major scientific advances were really the result of the consistent application of the methods of science. Decidedly not, he insists, for 'Everywhere science is enriched by unscientific methods and unscientific results while procedures which have often been regarded as essential parts of science are quietly suspended or circumvented.'[28] The success of science has been gained by the willingness of creative scientists to violate their own methodological rules, and to adopt without acknowledgement the methods and theories of others. The moral, for Feyerabend at least, is clear: 'there is only *one* principle that can be defended under *all* circumstances and in *all* stages of human development. It is the principle: anything goes.'[29]

The onus is clearly upon Feyerabend to identify 'the unscientific methods' he feels are so widely used in science. Among other examples he considers the case of Copernicus. In 1543, in his *De Revolutionibus*, Copernicus initiated the first scientific revolution of the modern period. The Ptolemaic cosmology of antiquity with its earth-centred universe was rejected by Copernicus and replaced with a system centred on the sun. Three charges are raised by Feyerabend against Copernicus's procedures.[30]

1   The heliocentric idea itself was 'stolen' from Philolaos, 'a muddleheaded Pythagorean' of the fifth century BC. While an idea's origin may be of some historical interest, its relevance to Feyerabend's argument remains obscure. Whether the idea came to Copernicus in a dream, from an ancient text, or from his own intellectual resources, can have no impact on the nature of the argument he developed in the text of his book.[31] Newton was frequently faced with the charge that others had anticipated his work. *That* idea, his rivals insisted, is *mine*, not *yours*. Shortly before he published *Principia* in 1687, he was informed that Hooke was claiming the discovery of the inverse square law. Newton's exasperated response contained a crucial distinction:

Now is this not very fine! Mathematicians that find out, settle, and do all the business, must content themselves with being nothing but dry calculators and drudges; and another, that does nothing but pretend and grasp at all things, must carry away all the invention, as well as those that were to follow him as well as those that went before.[32]

Science of course needs ideas; the 'business' of science, however, as Newton realised, lies just as much in the working out of those ideas as deeply and as precisely as possible. The idea behind the *De Revolutionibus* can be stated in a single sentence; its working out took a further 500 pages. The problem facing Copernicus was to provide geometrical constructions which would allow the known motions of the heavens to be modelled on the assumption that the universe was heliocentric, and that all such motions were ultimately reducible to purely circular motions. The question is a purely technical one and was so noted by Copernicus by the addition to the title-page of his book the warning: 'Let none enter who knows no geometry.'

2 Copernicus could only succeed in the introduction of his new scheme 'by violating reasonable methodological rules'. All experience shows us that the earth is motionless and that all other heavenly bodies orbit the earth. The only way to move away from this position is by allowing, in Galileo's phrase, 'reason to conquer sense'.[33] Feyerabend's argument requires him to exaggerate considerably the supposed implausibility of the Copernican hypothesis. In educated circles, at least, the possibility of a moving earth had been widely considered. Nor had it been dismissed out of hand as absurd. Oresme, for example, in the fourteenth century had offered a number of acute criticisms of the arguments offered for the immobility of the earth.[34] For the proposition was indeed argued for and not simply offered as an undeniable observational truth. Ptolemy himself had found it necessary to argue against those who thought it 'strange' that the earth was immobile, and accused them of relying too much on their own personal experience.[35] Hence the discussion, spanning centuries, on the fate of an arrow shot straight up into the air, and balls dropped from the masts of moving ships. The issue was, emphatically, an arguable one. Further, it was an issue arguable in terms of theories of motion. In short, it was a scientific issue and one that had troubled thinkers for centuries.

3 The reasons Copernicus offered in support of his views derive in part from the hermeticists and in part from 'a mystical faith in the fundamental character of circular motion'.[36] Copernicus argued for the central position of the sun in the following passage:

In the middle of everything stands the Sun. For in this most beautiful temple who could place this lamp in any other better place than one from which it can illuminate all other things at the same time? This Sun some people call appropriately the Light of the World, others its Soul or its Ruler. Trismegistus calls it the Visible God, Sophocles' Electra calls it the All-Seeing. Thus the Sun, sitting on its Royal Throne, guides the revolving family of the stars.[37]

Little here of rational argument.

The passage, however, carries a simpler meaning. Copernicus had to face the problem of why the sun rather than, say, Jupiter was at the centre of the universe. In the absence of any astronomical reasons he sought some modest support in analogy. The sun, consequently, is seen king-like ruling his orbiting planetary subjects. Having rapidly exhausted the resources of analogy, he turned to antiquity and indiscriminately selected such authorities as Hermes Trismegistus who had, however peripherally, spoken of the sovereignty of the sun. Copernicus was not led to his belief in a heliocentric universe by reading hermetic texts; rather, once having committed himself to such a view, he cast around for support of any kind.[38]

There is in fact no mystery concerning the sources of his ideas. They are identified with force and clarity in the preface to his *De Revolutionibus*. Not only was traditional astronomy, he complained, incapable of explaining such basic phenomena as the length of the year, but the system was unworkable. Individual parts may be adequate, but when combined in one system, as was seen in chapter 6 above, the result is a monstrous hybrid bearing no relation to reality. One strand of Copernican thought clearly aimed to dispense with the piecemeal approach to astronomy and to impose some systematic order on the subject.

As part of this system Copernicus added the further requirement that celestial bodies should be allowed to move only in circular orbits with uniform motion. For Feyerabend this is a sign of Copernicus's mystical faith. More reasonably, it can be seen as a legitimate research programme. As we saw in chapter 6, much of science, including science of the highest kind, consists of little more than the obsessive working out of the details and implications of an initial hypothesis. Copernicus with his commitment to uniform circular motions joins Newton and the notion of universal attraction, and Darwin and the principle of natural selection. Others before them had contemplated such notions and made initial and partial attempts to develop them further. Much of the success of Copernicus, Newton, and Darwin, intellect apart, can be traced to the obsessiveness with which they pursued their initial interests. Whether this was due to a 'mystical faith', ambition for a knighthood, or an unhappy childhood, is irrelevant to the nature of the arguments developed. As proof of this it merely needs to be noted that the *De Revolutionibus* can be fully comprehended by someone unaware of its supposedly incriminating preface. In the same manner many have gained an adequate understanding of the works of Newton and Darwin long before they knew anything of their lives, characters, and more idiosyncratic beliefs.

Is science uniquely effective?

Against the claim that science possesses a uniquely effective method Feyerabend has offered two main arguments. It is conceded that in many obvious areas science is extremely effective. Could there not be, however, other methods just as effective? There could indeed, Feyerabend has repeatedly insisted, and one such, mythical thinking, is accordingly identified. The inventors of myth, he argues:

> invented fire, and the means of keeping it. They domesticated animals, bred new types of plant . . . invented rotation of fields . . . crossed the oceans in vessels . . . and demonstrated a knowledge of navigation . . . Thus, if science is praised for its achievements, then myth must be praised a hundred times more fervently because *its* achievements were incomparably greater. The inventors of myth started *culture* while rationalists and scientists just *changed* it.[39]

Feyerabend clearly has in mind the so-called neolithic revolution, a term coined by the archaeologist Gordon Childe, and more recently given a greater currency in the writings of Claude Lévi-Strauss.[40]

Let it be conceded with Feyerabend and Lévi-Strauss that long before the emergence of writing man had mastered fire, domesticated animals, learnt how to grow crops, settled in villages and, most unlikely, 'demonstrated a knowledge of navigation'. Are these achievements comparable with the successes of science?

Only, in fact, if it could be shown that we are considering cases of invention and discovery in both cases. It is possible that the advances of the neolithic period were, like language, things which emerged. Just as, over long periods of time, and in numerous places, language developed, so too did agriculture and other neolithic traits. Did some one person or group invent, say, the future tense or the personal pronouns of the Indo-European languages? Neither language nor the domestication of animals arose in this manner. Consequently, although we can very roughly date their appearance, there are simply no inventors to be identified.

Science, in contrast, is more personal, more specific and singular, and also more locatable. Whatever methods they adopted, we can be reasonably sure that the proof of Pythagoras's theorem or the construction of a reliable manner to measure the earth's circumference, were the conscious achievements of, at most, a few individuals. It is not merely that in the one case we can claim, whether accurately or not, to identify in the figures of Pythagoras and Eratosthenes the actual inventors, and that in the other case we have only anonymous hordes. The argument is stronger and more explicit: the discoveries of science are and must be individual achievements; those of myth could never have been.

A further indication of the difference between myth, in Feyerabend's sense, and science, lies in the sheer multiplicity of neolithic revolutions. In marked contrast there has been only one scientific revolution. While people in Europe, Asia, America, and Africa were all separately and independently learning how to farm, to domesticate animals, to master fire, one small group of scholars in a relatively short time worked out the procedures and the details of axiomatic geometry. Their work has been duplicated by no one, and by no other culture. Consequently, the geometry which would be later taught in Padua, Paris, Cambridge and Baghdad was in all cases an alien import and can be traced directly back to a text of Euclid.

A further false suggestion is made in the claim that the inventors of myth 'demonstrated a knowledge of navigation'. Many communities have worked out various techniques to find their way around some familiar and localised areas. This, however, is very different from navigation. Navigation is concerned with the determination of latitude and longitude and was no more a part of the skills of the neolithic mariner than was an ability to fly. Early mariners may well have been able, within limited areas, to find their latitude, and to sail to many places. Yet no one, before the eighteenth century, could reliably and accurately locate their longitude. Numerous European shipwrecks prove the point. In 1707 Admiral Sir Clowdisley Shovell ran the British fleet on to the Gilstone Ledges in the Scilly Isles and died along with almost 2,000 sailors. The disaster was enough to convince the British government that reliable methods to determine longitude must be found. A prize of £20,000 was offered to whoever could determine longitude to half a degree or thirty miles. The problem remained unsolved until 1762, when John Harrison provided a sufficiently accurate marine chronometer.

Secondly, Feyerabend has argued that the apparent dominion claimed by science over its long-vanquished rivals is intellectually baseless. It is 'the result not of research, or argument, it is the result of political, institutional, and even military pressures.'[41] The fact that astrology and divination have been replaced by apparently more effective approaches to forecasting means nothing. Atomism, Feyerabend points out, after its rejection by Aristotle was virtually ignored for two millennia before it was revived in seventeenth-century Europe. Could there not be similar revivals in the future, with forecasters abandoning their computers for sheep's entrails, and physicians preferring an astrological diagnosis to biochemical analysis?

It is natural to object that surely these are battles fought and won by science long ago. But, Feyerabend insists, you will seek in vain for such confrontations. Instead, there are numerous accounts of how the powerful institutions of science, beginning with the Royal Society and Académie des sciences in the seventeenth century, were so able to define what counts as rational thought as to deny even a hearing to any alternative traditions. Science may well be all it claims to be, and astrology and divination by entrails may be quite useless; the claims are, however, according to Feyerabend, mere assumptions. They remain to be tested.

There is some truth in Feyerabend's argument. Having sought in the seventeenth century to distance themselves from the occult sciences, the founders of the Royal Society were in no mood to allow their rivals a fair hearing. The astrologers themselves went some considerable way to collect sufficient data to allow their theories to be evaluated. They even sought the patronage of the Royal Society for their 'experiments in astrology'. The society's secretary, Robert Hooke, dismissed astrology as vain and ignored the request.[42]

More recent examples are plentiful. The case of Immanuel Velikovsky is well known and has been fully documented by De Grazia.[43] The scientific establishment sought not merely to ignore the unorthodox cosmological speculations of Velikovsky, nor only to dismiss them out of hand and unread; they actually lobbied for their suppression. For centuries, though seldom as publicly, the institutions of science have preserved their authority by denying their rivals a fair hearing or adequate test.

So far there is little to disagree with in Feyerabend's position. It is now widely accepted that science, like any other organised activity, adopts institutional as well as intellectual strategies to maintain its position. We can therefore concede that the Royal Society has never seriously considered the efficacy of rain-dances to control the weather, and that they tacitly and without proof assume that 'scientific rationalism is preferable to alternative traditions.'

Feyerabend, however, is not content with merely pointing out the omissions of orthodox science. He goes on to advance the more contentious claim that there are elements to be found in the numerous alternative traditions which are certainly as good as, and may even be better than, any product of scientific rationalism. In China, for example, where competition between science and traditional medicine has been allowed, it has been shown that 'traditional medicine has methods of diagnosis and therapy that are superior to those of Western scientific medicine.' Similar results have been obtained from other societies. More generally, 'non-scientific ideologies, practices, theories, traditions can become powerful rivals and can reveal major shortcomings of science if only they are given a fair chance to compete.'[44]

The claim is easy to make but much harder to justify. No references are given for the certainly disputable claim that acupuncturists and herbalists have shown 'that they can diagnose and heal illnesses whose effects scientific medicine recognises but which it neither understands nor heals.'[45] The only way such claims can ever be made plausible is by the adoption of double standards. The claims of scientific medicine are subjected to the most rigorous and critical tests and are consequently found frequently to evaporate into a mixture of placebo effect, a natural tendency for bodies to heal themselves, and side effects more lethal than the initial illness. If the same standards were applied to the not untypical claim, taken from a catalogue advertising alternative medical texts, that primrose oil 'holds out new hope for sufferers of heart disease, arthritis, multiple sclerosis, menstrual and menopausal symptoms, schizophrenia and even cancer',[46] they would be likely to prove no less volatile.

Further illustrations can be found in almost any of the numerous works of Brian Inglis. In *Fringe Medicine* (1965), for example, he argued convincingly, on the basis of controlled trials and statistical analysis, that many new drugs are less effective and more dangerous than is generally allowed. Twenty years later few would reject his analysis. They would, however, find it less easy to accept his account of psychic surgery. 'In the Philippines the psychic surgeons do not use knives, claiming that they can enter the patients' bodies without them. This type of "surgery", too, is ordinarily painless, and leaves hardly a trace, even when a tumour, say, has been extracted.'[47] What does Inglis say to the objection that on examination the supposedly extracted human tumours prove to be animal entrails? He admits the fraud but continues to argue that patients are really cured. Perhaps, he speculates, 'some form of psychokinetic energy' closes the scarless incisions.

It is also misleading for Feyerabend to suggest that there are a number of alternative traditions, such as herbalism and astrology, waiting ignored and untested in the wings. Suffer from syphilis, meningitis, or a perforated appendix almost anywhere in the world, and the treatments offered will be remarkably similar. The same drugs, surgical techniques, and diagnostic procedures will be undertaken whether the patient is in Tokyo, Lagos, Paris, or New York. Visit herbalists and astrologers in the same areas, with any complaint, and diagnosis, treatment, and the very concepts used to describe the illness are likely to have nothing in common. Even within a single 'tradition' herbalists seem to offer little more than competing panaceas. Thorson's 1988 catalogue, for example, contains claims for vitamin E, lecithin, grapes, vegetable juice, and the already mentioned primrose oil. All are offered as cures for complaints ranging from piles to schizophrenia. And this is merely one slim, British catalogue.

Nor is anything clearer if we turn to astrology. The injunction to examine, analyse and test astrology evokes the response: which astrology? The astrology of the East or the West? If of the West, the classical astrology of Ptolemy or its later development? Or again, perhaps attention should be paid to the unorthodox thesis of James

Vogh in his *Arachne Rising* (1977) that the spider is the missing thirteenth sign of the zodiac. A discovery which, his blurb-writer informs the reader, 'clearly demands a reconsideration of both astrology and the paranormal'. One of the awkward features of alternative traditions is their sheer number. They also share a tendency towards endless fission.

No sooner, for example, have we learned that the claims of acupuncture are soundly based than we find the market invaded by 'Acupuncture without the Needles'. Acupressure, for the squeamish, apparently activates the same 'fundamental principles of Chinese medicine', but with thumb and forefinger manipulation in place of the traditional needles. Or consider iridology, the view that markings on the iris reveal the functional state of the various organs of the body. The discipline was founded by the Hungarian von Peczely in the 1860s, and can be studied in such texts as Bernard Jensen's *Iridology* (1982). A more critical survey of the field by R. Worrall revealed a bewildering variety of charts. While they agree in part – the leg area, for example, is usually linked with the 6 o'clock position – the nineteen charts consulted by Worrall varied considerably in both location and interpretation.[48]

Some systems are so novel that they can mislead even the very sharp Richard Feynman. While investigating alternative systems in California he settled in a pool where there was 'a beautiful girl sitting with a guy who did not seem to know her'. While thinking how he could get to know the girl, Feynman heard the man ask if he could give the girl a massage. She agreed. Feynman is full of admiration for what he takes to be 'a nifty line'. But, to his surprise, the man starts to rub the girl's big toe.

> 'I think I can feel it', he says. 'I feel a kind of dent – is that the pituitary?'
> I blurt out, 'You're a helluva long way from the pituitary, man!'
> They looked at me, horrified – I had blown my cover – and said, 'It's reflexology.'[49]

It is difficult to see how science can legitimately face its supposed competitors. Lacking any coherence, and presenting an immense multiplicity of competing views, only a minute sample could ever be seriously examined. The results of such an analysis would be unlikely to satisfy anyone. If a Royal Society committee ever did succeed in disproving, to the satisfaction of everyone, the efficacy of, for example, Nupe sand divination, their efforts would have established little. There would still remain several thousand unexamined divinatory systems, supported by adherents indifferent to the weaknesses of the Nupe system. Indeed, new systems emerge quicker than they could ever be comprehensively surveyed.

The only sane strategy for science to pursue is the examination of principles rather than systems. Any alternative system which violates such basic principles as the conservation of energy, or has information transmitted faster than light can, of course, be dismissed without further examination. Unfortunately few critics will be convinced by such an approach. The scientist's reluctance to consider, say, processes which violate the second law of thermodynamics will be taken by critics as simply one more sign of his arrogance and prejudice.

At least, on this matter, scientists can answer Feyerabend's objection. It is certainly true that orthodox science has failed to examine fairly and in detail many hundreds of alternative traditions. Even astrology, so important in the development of western astronomy, has yet to receive this recognition. On the surface it may thus appear, as Feyerabend has argued, that such alternative traditions are rejected not as a result of reasoned argument, but through political pressure. If, however, the scientist has argued for the truth of a number of deep principles, and if also these principles have been the subject of intense critical debate and repeated and increasingly sophisticated testing, and have survived the procedure, then the scientist is likely to reject out of hand any theories incompatible with these principles. To the casual observer it may appear outrageously dogmatic; to the scientist it is a common procedure and one in which he sees himself acting sensibly and rationally. After all, they apply these principles on a daily basis to each other's work. Why then should the research of others be exempt?

Scientists, of course, often get their arguments wrong, and appeal to principles which are more dubious than deep. As is well known, Lord Kelvin, the leading British physicist of his day, dismissed Darwin's work on the ground that it violated the principles of thermodynamics. The sun could be no more than 100 million years old; evolution demanded a much longer period in which to operate;

therefore evolution must be rejected. Kelvin wasted no time pursuing the minutiae of the geological and palaeontological evidence on which evolution was based. Physics in the guise of thermodynamics had spoken clearly and whatever failed to fit into its scheme had to be rejected. Kelvin's thermodynamics were later shown to be wrong. Unaware of radioactivity, he had inevitably failed to allow for its effects in his calculations.[50]

Such errors offer no consolation to Feyerabend. His argument has been that science reacts to its rivals with force and pressure, rather than with research and argument. In reality the arguments are present and the research results are reported even though, at times, the arguments may be invalid and the research inconclusive. As an example of a scientist querying the validity of an alternative tradition by an appeal to scientific principles we may take the reaction of Einstein to ESP. All known forces – gravity, strong, weak, and electromagnetism – decrease in power over distance. Telepathy, in contrast, seems independent of distance. Psychics can perform as readily across the Atlantic as they can over the dinner table. To a physicist like Einstein this was most surprising. Accordingly, he wrote to a correspondent in 1946, 'But I find it suspicious . . . that the distance of the subject from the cards or from the "sender" has no influence on the result. This is, a priori, improbable to the highest degree, consequently the result is doubtful.'[51]

Atomism

Having seen how one philosopher responded to the science of his day, and how another has responded to the very notion of science itself, we will next consider the reactions of many philosophers over the centuries to the single topic of atomism.

Origins

Atomism began as philosophical speculation by Democritus and Leucippus. They were in fact responding to the major intellectual crisis of their time, precipitated by the Eleatics. The arguments of one Eleatic, Zeno, concerning the impossibility of Achilles ever overtaking the tortoise, are with us today and, although over 2,000 years old, can still niggle even expert mathematicians and logicians. Zeno, however, had been preceded by Parmenides. In a quite abstract manner Parmenides had argued that Being, whatever exists, is eternal and unchangeable. Nothing, therefore, seemed predicable of Being other than that *It is*. Even motion was impossible, with Being forever 'remaining in the same place'. The well-known antinomies of motion propounded by Zeno reinforced the Parmenidean claim. Whereas Parmenides had argued for his austere concept of Being directly, Zeno had shown that any alternative view, that bodies moved, for example, led directly and fairly rapidly to a number of intractable paradoxes.

The philosophers of the fifth century sought to respond with more realistic accounts of the world, accounts which allowed for change and motion. At the same time they felt the constraints imposed upon them by the Eleatics. How could one thing change into another? How can water change into ice? How could such a thing come to be? Such processes, for Parmenides, required the absurd assumption that Being could come from Non-being, what *is* from what *is not*. One possible solution was to accept the Eleatic viewpoint but to rewrite the normal notions of change and becoming. Let matter be composed of numerous minute particles, eternal and unchangeable. As the particles are eternal and changeless, they can never have come to be. Change in the world can now be seen as regroupings of changeless particles. Once the initial assumption was made further elaborations could follow. The fullest account of ancient atomism we possess is to be found in the *De rerum natura* of Lucretius from the first century BC.[52]

By this time the threat posed by the Eleatics had receded. Atomism needed independent justification. Lucretius offered in its defence a complex argument. He began by appealing to two general but speculative truths:

1 Nothing can ever be created by divine power out of nothing.
2 Nature never reduces anything to nothing.

In their defence, despite numerous rhetorical flourishes, Lucretius has nothing more plausible to offer than the claim that it would be absurd to deny them.

Given that matter was conserved, Lucretius concluded that there must be 'bodies whose existence you must acknowledge though they cannot be seen'. He was, however, well aware of the objection that invisible particles were indistinguishable from fictitious particles. In defence he pointed to a number of simple phenomena best understood as the gradual loss of imperceptible atoms. Wet clothes do dry, ploughshares do wear away, and even massive sea cliffs – 'exposed to prolonged erosion by the mordant brine' – can erode. To sum up:

Whatever is added to things gradually by nature and the passage of days, causing cumulative increase, eludes the most attentive scrutiny of our eyes. Conversely, you cannot see what objects lose by wastage of age . . . or at what time the loss occurs. It follows that nature works through the agency of invisible bodies.[53]

Lucretius realised that for atomism to be acceptable he must also show that 'there is a vacuity in things.' Once more the arguments were mainly derived from familiar phenomena. Water seeps through rocks into caves, noises pass through walls, and cold penetrates into bones. None of these phenomena would be possible, Lucretius concluded, unless there were vacancies in bodies.

Further, some things outweigh others of equal volume. If we allow that there is as much matter in a ball of wool as in one of lead, then it should weigh as heavily. Unless, of course, Lucretius argued, wool has more vacuum in it than lead.[54]

Lucretius was thus able to conclude that nature consists of just two things: 'bodies and vacant space in which the bodies are situated and through which they move in different directions'. At this point he turned to the issue of the nature of the atoms whose existence he had established, and thereafter he sought to show how all aspects of life could be explained with their aid. The processes of vision, hearing, smell and taste, as well as sleep, dreams, sexual attraction, are all shown along with such meteorological phenomena as thunder and lightning, to be basically atomic phenomena.

The Aristotelian response

The first major philosophical response came from Aristotle. Unfortunately for the fate of atomism he succeeded in producing a neat little dilemma which successfully persuaded generations of later philosophers and scientists that atomism was incoherent and should not be taken seriously. Aristotle had argued that 'nothing that is continuous can be composed of indivisibles.'[55] A line, for example, cannot be composed of points. For it to be so, the points would have to be in contact with each other. To achieve this, either 'whole is in contact with whole, or part with part, or part with whole'. As, by definition, points and atoms are indivisible and cannot have parts, continuity can only be gained by allowing wholes to be in contact with wholes. But, Aristotle emphasised, to say two wholes are in contact with each other at every point is to say that we have one point and not two.

The damage was done and throughout the medieval period whenever atomism was considered this objection of Aristotle would be produced. We find William of Alnwick (c. 1270–1333) arguing:

If a continuum were composed of indivisibles . . . the addition of indivisible to indivisible would cause an increase in size . . . to create an increase in size it would be necessary that one indivisible touch another . . . either according to their wholes or according to a part. Not according to a part, since [an indivisible] does not have a part. If according to their wholes, then no increase in size will occur, since whole touching whole causes no increase in extension.[56]

Further objections were raised. They usually took the form of pointing to the absurd or contradictory consequences which would follow from the atomic hypothesis. Al Ghazzali (1058–1111) neatly argued that the revolution of a wheel showed the impossibility of atomism. For, does not the circumference move a greater distance than parts at the centre? Allow then the wheel's rim to move just the breadth of one atom. What of the central parts? 'they will either move less than an atom and hence an atom will be divided, or they will not move at all and it will follow that all parts of the wheel are separate so that some move while others do not, but remain at rest. But our senses tell us this is false.'[57]

Whether because of the influence of Aristotle or not, there were very few atomists throughout the medieval period.[58] There were, no doubt, other reasons for atomism's unpopularity. For some atomism was too close to materialism and atheism to be attractive. In 1348

Nicholas of Autrecourt was condemned by the Paris faculty for maintaining that there was nothing in nature other than 'the motion of the combination and separation of atoms'.[59] Thomas Hobbes, too, had to face the charge that his supposed atomism had led him into atheism.

Nor was atomism free from scientific objections. It was one thing to present atomism as a plausible explanation of the phenomena of the kitchen and the farm; it was another to argue seriously for the existence of homogeneous, invisible, and indivisible atoms. A little saffron, Bacon noted, will cover a whole hogshead of water, thereby showing that there is 'in things a much more subtle distribution and comminution' than was apparent, while a little civet 'will infect a suite of two or three large rooms with its odour'. Yet, plausible as such accounts may be, there remained genuine difficulties. What, Bacon asked, about contraction and expansion? If atoms were solid, indivisible, and without any interspersed vacua, how could bodies ever be compressed? But once we allow atoms to be composed of matter and space they lose their indivisibility and homogeneity. Further, many scholars were so opposed to the notion of the void that, rather than accept its existence, they would willingly reject all atomic notions as well.

First revival

When atomism was revived in the seventeenth century it was seized upon at once by scientists as a new and fertile mode of explanation. One of the earliest atomists was the English mathematician and physicist Thomas Hariot (1560–1621). He informed Kepler in 1606 that atomism could be used to explain why light is partially reflected and partially refracted from the surface of a transparent medium. As the surface is not really continuous, Hariot argued, but composed of 'corporeal parts which resist the rays', and 'vacua which the rays penetrate', refraction can be seen to be 'nothing else than an internal reflection'. Hariot concluded his letter to Kepler by advising him that, if he wished to discover the mysteries lying within 'nature's house', he should 'abstract and contract' himself into an atom.[60]

Later in the century Robert Boyle and Newton based much of their work on the atomic hypothesis. Thus, in *Principia* (Book II, Prop.

XXIII), Newton set out to show that a form of Boyle's law could be derived from the assumption that a fluid was composed of 'particles fleeing from each other, with forces that are inversely proportional to the distances of their centres'. He went on to emphasise the importance of the assumption by laying it down as Rule 3 of the Rules of Philosophical Reasoning which preface Book III:

The extension, hardness, impenetrability, mobility and inertia of the whole, result from the extension, hardness, impenetrability, mobility, and inertia of the parts; and hence we conclude the least particles of all bodies to be also all extended, and hard, and impenetrable, and movable, and endowed with their proper inertia. And this is the foundation of all philosophy.

At a more speculative level Newton wrote about the possibility of forces operating between the particles to produce various chemical reactions. In this manner, just as the sun attracts the earth by the force of gravity, there is a comparable specific force operating, for example, when 'spirit of vitriol' is mixed with 'Sal Alkali'.

Despite the obvious appeal to working scientists of atomism, many philosophers, on purely philosophical grounds, found it unacceptable. Descartes, for example, was forced to deny the existence of indivisible atoms. After arguing that the essence of matter is extension, it followed that if atoms, 'exist, they must necessarily be extended, however small they are imagined to be; so we can still divide any one of them in thought . . . There is nothing we can divide in thought but we can see to *be* divisible.'[61]

In a similar manner, though considerably later, Leibniz drew the inevitable philosophical conclusion that atoms could not possibly exist. Central to Leibniz's metaphysics is the principle of sufficient reason, through which he claimed that God's perfection required: 'that all his actions should be agreeable to his wisdom; and that it may not be said of him, that he has acted without reason; or even that he has prefer'd a weaker reason before a stronger.'[62] In the hands of Voltaire the principle became 'Everything is for the best in the best of all possible worlds', and inspired him to write *Candide*. Leibniz, however, was led to derive the secondary principle of the identity of indiscernibles. God could not create, he argued, 'two real absolute beings, indiscernible from each other'. For, if he had, nature would be forced to act without reason in placing one object at one place and

the other elsewhere. Consequently, there cannot be indistinguishable objects such as atoms.

Arguments of this kind proved too sectarian to be generally
convincing. They were too much a part of a specific metaphysics to
appeal to anyone other than Cartesians and Leibnizians themselves.
Consequently someone like Newton, who was neither, could feel free
to ignore them. More effective were various general objections, made
over the centuries – objections which come as readily to the systematic philosopher as they do to the theologian, scientist, and man in the
street. The most persistent complaint emphasised that however
useful atoms may be to the scientist they are, at best, theoretical
entities. Many have felt distinctly uncomfortable at the ease with
which atomists can conjure explanations and entities to solve any
problem placed before them. Why are honey and milk pleasant to
taste, while wormwood is so disagreeable? Because, Lucretius
claimed, the former substances are composed of 'smooth round
atoms' and the latter of 'hooked particles'. Suggestions of this kind
come more from the imagination than from reason. Consequently, in
the absence of stronger evidence, many were unwilling to recognise
atoms as anything other than hypothetical fictions.[63]

A second major worry has been the suspicion that atomism is
essentially a very limited view. If all bodies are no more than a
collection of atoms, how can we distinguish between living and
non-living bodies? The physiologist William Harvey, discoverer of
the circulation of the blood, was one who stressed the point. A house
could perhaps be understood in terms of the bricks, timber and other
things going into its construction. But, he insisted, 'There is a greater
and more divine mystery in the generation of animals than the simple
collecting together, alteration and composition of a whole out of the
parts would seem to imply.'[64]

Atomism, however, faded from the centre of intellectual interest
rather quickly. This had little to do with the philosophical objections
of Descartes, Leibniz, or anyone else. Instead, interests shifted,
alternative schemes emerged, and atomism rather than being rejected appeared merely pointless and irrelevant. Consequently, it was
ignored to such an extent that the English translation of Macquer's
influential *Dictionary of Chemistry* (1771) lacked altogether entries on
Atoms and Corpuscles.

The new language of science dealt more with such topics as *force*, *activity*, *power* and *affinity*, and its roots lay in the work of Newton and Locke. Newton's universe was dominated by gravitational attraction at the macroscopic level, and by as yet unidentified forces of attraction and repulsion at the microscopic level. Locke, too, had spoken about powers. There were for Locke three sorts of qualities in bodies. The primary qualities such as 'bulk, figure, number', are really in the bodies whether we perceive them or not; the secondary qualities – colours, sounds, smells – were produced in us by the primary qualities, but were not actually present in the bodies themselves. In addition there were powers in bodies, the result of the particular arrangement of the primary qualities, capable of changing 'the bulk, figure, texture and motion of another body, as to make it operate on our senses differently from what it did before. Thus the sun has the power to make wax white, and fire, to make lead fluid.'[65]

Having seen Newton's success in working out the manner in which one power or force, that of gravity, operated within the universe, ambitious scientists sought to emulate his achievements in other areas. In chemistry, for example, throughout much of the eighteenth century talk of atoms simply disappeared. Instead, chemists tended to speak of how one substance could have an *affinity* for another. Thus, Macquer's *Dictionary* required eighteen pages to present the theory of affinity, but had nothing to say about atomism. Philosophical backing for these views soon appeared in the critical philosophy of Kant.

Some of Kant's views had been anticipated by the Jesuit, Roger Boscovich, in his *Theoria naturalis philosophiae* (1762).[66] The hard impenetrable particles referred to by Newton, Boscovich argued, could not account for elastic collision. Normal objects, like tennis balls hitting a wall, rebound as a result of elastic deformation. But the atoms of Newton cannot be deformed. Therefore, when two atoms collide, contact and rebound must be instantaneous. Thus, absurdly, atoms must have two velocities at the same time. Boscovich proposed to solve the problem of material atoms by retaining the atoms, but denying that they could be material. Atoms, he insisted, were simply mathematical points surrounded by attractive and repulsive forces. The point atoms of Boscovich, it turned out, surprisingly, were no less explanatory than the hard, impenetrable atoms of Newton.

Kant was led to a similar position. Substance, as described by Locke, was dismissed as unknowable, something forever hidden from human experience. We were led to the notion of substance, Kant argued in his *Critique of Pure Reason* (1781), through considerations of causation and force: 'This conception of causality leads us to the conception of action; that of action to the conception of force; and through it, to the conception of substance.'[67] Further, he declared, 'Where action (consequently activity and force) exists, substance must also exist, and in it alone must be sought the seat of that fruitful source of phenomena.'[68] We come to know substances in space through the forces, attractive and repulsive, we encounter. When Kant came, in his *Metaphysical Foundations of Natural Science* (1786), to apply the critical philosophy to physics, Proposition 1 of the treatise stated:

> Matter fills space, not by its pure existence but by its special active force.

This notion of the universe as a plenum filled with forces rather than things was introduced to English readers through the writings of the poet Coleridge. It was soon found echoed in the thought of two leading British scientists, Davy and Faraday. Both found the ideas of Boscovich and Kant better suited to their scientific work than the atoms of Boyle and Newton.

Here, for example, is Faraday struggling to make sense of electrical conductivity in terms of traditional atomism. For the atomist, matter consisted of atoms separated from each other by empty space. Thus, the only continuous part of a rod of shellac, as of a rod of platinum, would be empty space. It would therefore appear, Faraday argued, that the insulating properties of shellac, as also the conducting properties of platinum, were both caused by space. But 'If space be a conductor, how then can shell-lac, sulphur, etc insulate? For space permeates them in every direction. Or if space be an insulator, how can a metal or other similar body conduct?'[69] In reply Faraday turned to 'the atoms of Boscovich . . . mere centres of forces or powers, not particles of matter'.

Second revival

If the physicists had little need for atoms, what then of the chemists? There would at first sight seem to be a stronger call for atomism from working chemists, struggling to make sense of chemical reactions, than from other scientists. And, indeed, the nineteenth-century revival of atomism was largely the creation of chemists. It began with John Dalton, an English chemist, pondering over a problem which could never have troubled his predecessors, Boyle and Newton.

One of the great achievements of eighteenth-century chemists had been the discovery that the atmosphere was not a homogeneous medium, but a compound mixture of oxygen, nitrogen, water vapour and, perhaps, more besides.[70] But, as long as it was accepted that the atmosphere was of uniform composition, questions about its stability and integrity could never arise. However, once accept with Dalton that it was composed of three or more elastic fluids with different densities, and at once chemists have a problem. Why did the fluids not form separate layers, with the densest gas at the bottom and the rarest at the top? As a good Newtonian, Dalton turned to *Principia*, where he read that the atmosphere was composed of 'small particles or atoms of matter which repel each other'.[71] Dalton added the further crucial hypothesis that 'The atoms of one kind did *not* repel the atoms of another kind, but only those of their own kind.'[72] This would allow, he concluded triumphantly, 'for the diffusion of one gas through another, whatever might be their specific gravities'.

It remained for Dalton to make his most original contribution to atomic theory. Again, this arose from consideration of a specific scientific problem. Traditional atomists had tended to concentrate on the size and shape of atoms. This, however, was of little help to Dalton when he was considering the question: 'why does not water admit its bulk of every kind of gas alike?'[73] Why, for example, would water admit greater quantities of nitrous oxide than nitrogen or hydrogen? Because, he speculated, the reaction depended upon 'the weight and number of the ultimate particles of the several gases', with the lightest being least absorbable. This was sufficient to get Dalton thinking of atomic weights, a more manageable and fertile notion than atomic shape. Given a few further assumptions Dalton could reach such conclusions as that the relative weights of the elements oxygen and hydrogen are in a ratio of 7:1, and on this basis begin to work out the basic laws of chemical composition.[74]

Yet even among chemists atomism met with considerable resistance. There remained numerous scholars, chemists and philosophers alike, who simply could not bring themselves to accept the existence of imperceptible, indivisible particles. This was not science, not the result of chemical experiment, William Whewell objected in his *Philosophy of the Inductive Science* (1840), but metaphysics:

> But if the atomic theory be put forwards ... as asserting that chemical elements are really composed of atoms, that is, of such particles not further divisible, we cannot avoid remarking that for such a conclusion, chemical research has not afforded, nor can afford any satisfactory evidence whatever.[75]

Equally, the great chemist F. von Kekulé could insist in 1867: 'The question whether atoms exist or not has but little significance from the chemical point of view; its discussion belongs rather to metaphysics.'[76]

One reason for the apparent indifference of chemists to atomism lay in the presence of an alternative means to describe chemical reactions, namely, the language of chemical equivalents. Textbooks of the period, for example, presented tables of equivalents, explaining that 'Chemical equivalents merely represent the relative quantities of substances which combine together, as referred to some standard of comparison.'[77] Or, as expressed in a later science dictionary: 'Bodies are said to be *equivalent* when they can be substituted for each other in combination.' For example:

> By experiments ... it has been shown that different but definite weights of the various metals are capable of displacing each other ... 100 parts by weight of mercury, 31.7 of copper, 32.5 of zinc, and 1 of hydrogen are each in a condition to supply the place of the other in combination with 35.5 parts of chlorine.[78]

Thus, while Dalton and other atomists would wish to claim that water, a binary compound, was made from one atom of hydrogen, and one seven-times-heavier oxygen atom, the less committed would simply see seven units by weight of oxygen combining with one unit of hydrogen to produce one unit of water. Perhaps, for convenience, they would speak of atoms of hydrogen and oxygen; when pressed, however, they would insist that this was simply shorthand for more accurate talk about chemical equivalents.

Final triumph

As the twentieth century began, neither physicists nor chemists were solidly united in their support for atomism. Friedrich Ostwald, winner of the 1909 Nobel chemistry prize, told the Faraday Society in 1904 that the atomic hypothesis was 'unnecessary'.[79] Positivist objections also still circulated widely. Thus, Ernst Mach in 1897 could continue to insist that atoms were only 'a mathematical model for facilitating the mental reproduction of facts'.[80]

Yet the denouement to the long story of atomism was at hand and it was delivered, not by philosophers but, as could have been expected, by scientists. The issue revolved around the phenomena of Brownian motion, first described by the Edinburgh botanist in 1827. Brown had noticed that when pollen grains about 1/4000 inches in length were suspended in water, they not only moved in the fluid, they also often changed form: 'A contraction or curvature taking place repeatedly about the middle of one side, accompanied by a corresponding swelling or convexity on the opposite side of the particle. In a few instances the particle was seen to turn on its longer axis.'[81] Prolonged observation convinced Brown that these strange movements arose from the particle itself and not from currents in the fluid. Nor could it be explained by supposing the particles to be animated. The same results could be obtained by suspending equally small inorganic particles in a fluid.

Brown had in fact witnessed the effect of molecular bombardment on the pollen grains. The demonstration of this point, however, took another eighty years. Early response to this proposal was dismissive. As well suppose, J. H. Conybeare of Oxford argued in 1828, the molecules to be inhabited, and to have philosophers among their population.[82] Others subjected the particles to acid and alkaline solutions, to light and darkness, magnetism and electricity, heat and cold, red and blue light, and much more besides. In so far as there was any consensus on the issue, it seemed to favour the notion that exceedingly small currents had been set up in the liquid. Others saw the source of the effects as electrical.

The issue was finally resolved by the French physicist Jean Perrin in a series of papers published between 1908 and 1911. Perrin approached the problem in a number of ways. One of the more important was concerned with the distribution of the granules. Under the molecular hypothesis, while the granules should be uniformly distributed in the horizontal plane, there should be an increasing concentration with depth in the vertical plane. The precise amount of variation was worked out and was finally confirmed after several years' careful and detailed experimental studies. Perrin's work also confirmed some theoretical calculations made earlier by Einstein.[83]

Perrin himself was in no doubt about the implications of his work. When in 1912 he delivered a paper on 'Les preuves de la réalité moléculaire' to the first Solvay Congress of Physics, it met with no real opposition. At last, the mathematician Henri Poincaré noted in 1912, 'Atoms are no longer a useful fiction; things seem to us in favour of saying that we see them since we know how to count them . . . In the procedures deriving from Brownian movement . . . The atom of the chemist is now a reality.'[84] Perrin presented the results of his work to a wider public in his *Les atomes* (1913). J. D. Bernal later recalled how as a young student he had read the book and for him, as for Poincaré before, Perrin's work 'demonstrated the existence of atoms and the possibility of counting them'.[85]

Thereafter the only scientific dissent came from such totally immovable positivists as Mach. Recent experimental successes, he insisted in 1910, though numerous and useful, proved nothing concerning atomic existence.[86] Something of Mach's wariness has persisted and could be found expressed in the work of his intellectual descendants, the logical positivists. For the positivist all expressions, other than logical constants, must either refer directly to specific sense-data, or be explicitly definable in terms of expressions which do refer directly to specific sense-data. Otherwise the expressions are simply meaningless. Consequently, as Ayer noted in 1936, several of his positivist colleagues dismissed the terms 'atom', 'molecule', and 'electron' as illegitimate. The tradition is continued in the writings of empiricists of a later generation such as Bas C. Van Fraasen.[87]

Conclusion

Within this long and varied history no single or constant pattern seems to be present. While atomism may sometimes have been favoured by both philosophers and scientists, at other times the theory has been accepted by one group and dismissed by the other. What, then, of the question, 'Is atomism a scientific or philosophical theory?' The question could be answered in a rather boring way by responding that while, for example, it was accepted for scientific reasons by Perrin, it was denied on philosophical grounds by Leibniz, and to continue accordingly throughout the histories of science and philosophy.

The more interesting point is that the issue has not always turned upon any simple observation–speculation distinction. Obviously the claim that magnesium has an atomic weight of 24.305 is no speculative truth. Yet what could be more conjectural than the assumption of Dalton that we can infer from the relative weights of bodies the relative weights of their ultimate particles? And what could be more observationally based than the simple analogies of Lucretius? In this field, as in others, it appears to have been the scientists who have more frequently been seduced by the licence of conjecture than the more sceptical philosophers.

9

CAN SCIENCE AND PHILOSOPHY CONFLICT?

Science is a surer path to God than religion.

Paul Davies

Philosophers seldom challenge the propositions of science directly. For to do so, to dispute over such issues as the atomic weight of oxygen, or the melting point of bismuth, they would have to become scientists themselves. How, other than on scientific grounds, can anyone evaluate the claim that the region on human chromosome 21, 21q22.1, really does code for the enzyme phosphoribosylglycinamide synthetase? There are, however, less central areas where philosophers have felt free to challenge a number of claims made by scientists.

Reluctant to restrict their work to the facts of the textbooks, scientists are as prone as any other group to comment on the political, social, and economic questions of the day. Their judgements on these issues, whether trivial, crass, or acute, are quite independent of their scientific work. Frederick Soddy, for example, winner of the 1921 Nobel chemistry prize for his work on isotopes, virtually abandoned science after 1920, preferring to devote the latter part of his life to questions of economics. Soddy was convinced that money was the supreme enemy of economic freedom and argued his views at considerable length in a number of books.[1] Such views were clearly in no way connected with his demonstration in 1903 that helium is a decay product of radium. Nor was Einstein's Zionism a consequence of his theory of relativity, or Dalton's atomism a function of his Quakerism.[2]

Whereas the domains of isotopes and Zionism are unlikely to be confused, other intellectual areas may be less distinct. One site of regular confusion is to be found at the boundaries of science. Intellectual borders are as troublesome as any political frontier and, like them, are likely to be transgressed in a number of ways. Consequently, the frontiers of the intellect often need as regular policing as any political boundary. Traditionally the task has been performed, at least in part, by philosophers. Two common forms of transgression have particularly attracted their attention.

The first temptation is to take the results of science and attempt to show how they can be used to establish conclusions about other areas of life.[3] 'Science is a surer path to God than religion', the theoretical physicist Paul Davies has emphasised. In this claim he is doing no more than repeating that of the theologian Robert Jenkin, who in 1700, conscious of the work of his great contemporary, Isaac Newton, could claim that 'infidelity could never be more inexcusable than in the present age.'[4] Many scientists since have argued that science demonstrates the existence of God.

Attempts of this kind to transcend the agreed boundaries of science are not uncommon. Evolutionary theory, for example, has led many to infer quite specific conclusions about the nature of morality, and the course of our future development. They inevitably invite philosophical comment. Some of these issues will be considered in greater detail below.

A second difficulty arises over the borders of science themselves and their exact location. Like most boundaries they are changeable and far from static, expanding sometimes to allow the inclusion of newly developed disciplines, and even, occasionally, contracting to expel no longer welcome subjects. Astrophysics was born in the nineteenth century once physicists had learned how to extend their techniques to the internal constitution of the stars. Similar processes have taken place more recently with the creation of biochemistry, neurophysiology and, among many other disciplines, molecular biology.

Not all attempts, however, to incorporate new material within the domain of science are as straightforward. Such is the prestige of science that many scholars have sought to locate their own disciplines well within the boundaries of science. The objection is often made, mainly by philosophers, that many such attempts are misguided. Kant, for example, in his *Metaphysical Elements of Natural Science* (1786), maintained that empirical psychology must always remain 'removed from the rank of a science proper'. As mathematics could not be applied to the phenomena of inner sense, and as mental acts cannot 'be kept separate and recombined at pleasure' like the elements of chemistry, Kant concluded that psychology could never offer anything more than a 'description' of the mind.[5] More recently Karl Popper has argued that 'there can be no prediction of the course of human history by scientific or any other rational methods.'[6] Locke insisted that 'a perfect *science* of natural bodies' was 'so far from being capable of any such thing, that I conclude it a lost labour to seek it'.[7]

Conversely, other philosophers have claimed to see the emergence of still unrealised sciences. Austin in 1956 looked forward to the creation of a 'comprehensive science of language',[8] while Locke included in his division of the sciences semiotics or 'the doctrine of signs'.[9] A little later Hume, in the preface to his *Treatise*, argued for the necessity of a 'science of man' to supplement the science of nature developed a century earlier by Bacon and others. John Stuart Mill, also, in his *A System of Logic*, attempted to show how there could be 'moral sciences' which included a science of human nature, history, ethology or 'the science of the formation of character', and sociology. Mill was aware that it was 'a common notion . . . that the thoughts, feelings, and actions of sentient beings are not the subject of science', but thought the objection misplaced.[10]

In more recent times several eminent scientists, having accepted the reality of psychic phenomena, have campaigned vigorously for their study to be accepted as a legitimate part of science. Judgements of this kind, about what is and what is not an acceptable part of science, need not themselves be strictly scientific issues. Consequently, they invite philosophical scrutiny. However, before considering the modern obsession with the psychic, the earlier case of phrenology will be examined. It will be seen here that despite a strong early challenge it was eventually decided that phrenology had no place within science, and that part of the critical case against its credentials was mounted by philosophers of the period.

Phrenology

Tell me where is fancy bred,
Or in the heart or in the head?

Shakespeare's question had puzzled more than poets long before anatomists in the late nineteenth century developed reliable techniques for its answer. Before this, sites of mental activity were located on the basis of superficial anatomical knowledge, crude analogies, and philosophical presuppositions. Aristotle, for example, dismissed the view that the brain was the site of sensation on the ground that 'when it is touched, no sensation is produced.'[11] It was plain to him that an organ, unitary, central, and the first to receive blood – the heart – 'must be the primary source of sensation'.[12] In contrast, Plato offered a tripartite division of the soul into reason, emotion, and appetite. In the *Timaeus* a crude and arbitrary attempt was made to locate the three faculties, with reason being assigned to the head, emotions to the chest, and appetites to the abdomen.[13]

A more widespread and persistent scheme favoured the three cerebral ventricles as the seat of the soul with, according to one traditional account, the first ventricle devoted to sensation, and the other two to reason and memory. Dating back to Herophilus in the third century BC, the scheme still had considerable support well into the sixteenth century.

An alternative tradition, beginning with Descartes, insisted that as the soul was immaterial and without extension it made no sense to speak of its location or its parts. Yet not even Descartes was bold enough to sever all physical connection between mind and body. Sometimes he spoke of the soul being united to all parts of the body conjointly, insisting that cutting off part of the body would not reduce the size of the soul. He also claimed for the pineal gland, located in the centre of the brain, a somewhat privileged position. Struck by its central position and its unitary structure among numerous paired and symmetrical parts, he saw it as the most likely site for the harmonisation of the two images reaching the brain from the two eyes, along with the impressions arriving from other paired organs.[14]

Others were less convinced. Thomas Willis, for example, argued in his *Cerebri anatome* (1664) that as the pineal body was found in many lower animals it could not be the seat of the rational soul. He preferred the corpus callosum, the link between the two cerebral hemispheres, as a more likely site. Several other choices were equally possible and were indeed nominated by rival scholars of the period. Clearly the issue was not one capable of resolution by the concepts and techniques available to seventeenth-century scientists and philosophers.

Further, there were arguments suggesting that the union between mind and body was not something sited at a single central point. Robert Whytt (1714–66), an Edinburgh physician, was so struck by the ability of decapitated vipers to continue to respond to stimuli for several days that he concluded the soul was coextensive with the body.[15] The view was rejected by Haller, the leading physiologist of the century, in his *Elementa physiologiae* (1757–66). Yet Haller felt unable to return to the crudities of Descartes and his successors. The mind, he insisted, could be identified with nothing less than the entire brain. Localisation as an issue had lost its appeal; it appeared to be entirely arbitrary and was without any kind of clinical support.

The issue was revived by Franz Gall,[16] a Viennese physician, early in the nineteenth century. While earlier students had aimed merely to locate broad faculties of reason and memory, Gall set out to identify all mental traits, including such detailed features as a sense of humour, aggressiveness, musical ability, and benevolence. He began with a number of inspired assumptions:

1   The brain is the organ of the mind.
2   The brain is a complex organ with each of its parts related in a unique manner to a particular mental faculty. More than thirty distinct parts and faculties are recognised.
3   The size of the organ is a measure of its power.
4   The size can be deduced from surface irregularities, bumps, and hollows of the skull.[17]

At once the mind was accessible to all, to both professional systematisation and popular dissemination. By some, such as George Combe, an Edinburgh lawyer, Gall was greeted as 'the Newton of phrenology'. Yet, despite initial enthusiasm for the discipline, and despite continued support throughout the century from a number of sensible and talented workers, phrenology was never really admitted into the scientific domain. Why did it fail? What did it lack?

There were, first, a number of serious scientific objections. The French anatomist Pierre Flourens, working mainly with pigeons, was able to recognise different functions for cerebrum (intelligence), cerebellum (co-ordination), and the medulla oblongata (respiration), but could find no further localisation within those areas. Anatomists also challenged the basic assumptions of phrenology. What, they asked, was the basis for the correlation between skull and brain? Where were the supposed thirty-three organs of the brain actually sited?[18] The Edinburgh anatomist John Barclay pointed out that examination of the brain revealed nothing more than 'A number of almost similar convolutions all composed of cineritous and medullary substance, and all exhibiting as little difference in their form and structure as the convolutions of the intestine.'[19] The neat lines carefully marking one area from another on the familiar phreno-logical busts were dismissed as so much 'creative fancy'.

There were also a number of strong philosophical objections, originating mainly from the so-called Scottish common-sense philosophers such as Thomas Reid. One feature of their thought was a deep distrust of hypotheses, conjectures, and analogical reasoning, a distrust they claimed to have derived from Newton. While it was true, they conceded, that each bodily function was served by its own part, the claim that minds operated similarly was based on no more than analogy. When men account for nature by appealing to the operation of various causes, Reid had argued repeatedly, they should be able to show that such causes really exist, and not simply appeal to a more or less plausible analogy.

Phrenology came to be dismissed in precisely these Reidian terms. The influential piece by Peter Roget, secretary of the Royal Society, on cranioscopy, contributed to the *Encyclopaedia Britannica* (1824), concentrated on just such methodological issues. Analogy, Roget insisted, was not proof. The claim had been made, he reported, that because bodily secretions are produced by different glands, particu-lar thoughts and emotions must come from distinct parts of the brain. But what of the stomach, Roget retorted: though it digested different foods it was not seriously proposed that 'one portion of the organ is destined for the digestion of meat, and another for vegetable matter.'

Further objections were raised by the German philosopher F. A.

Lange in his influential *The History of Materialism* (1865). While anatomists had reported their failure to find within the brain any of the distinct organs of phrenology, Lange raised the deeper point of whether it could ever be meaningful to speak of distinct organs or faculties of this type. Could there ever be, he asked, distinct organs for such attributes as benevolence or amativeness? I hear a child crying and rush immediately to its aid because of my well-developed organ of benevolence. But what has happened here? I heard the child cry, responded sympathetically, decided to act in a certain way, and finally carried out a complex series of movements. But to attribute all this to a single organ, Lange argued, would be 'to conceive the activity of the organ as that of the entire man':

In the organ of benevolence everything must concur; not only thinking, feeling, and willing, but also hearing and seeing. If I renounce this anthropomorphism, which only postpones the matter to be explained, nothing can appear to me more probable than that in the phenomenon under consideration my whole brain was engaged, although in very various degrees of activity.[20]

That is, to *be* benevolent is not simply a matter of having certain feelings, whether or not they are linked to a particular part of the brain. Also involved are the abilities to recognise certain situations, to analyse them, to be able to work out how best to respond to them, and the determination to act in a certain manner. Either these additional features are part of the organ of benevolence, or they are not. If they are not, then it is hard to see how we can be talking about an organ of *benevolence*, and if sensation, perception, cognition, and volition are all part of it, then, in turn, it is equally difficult to see how we can be talking about a mere *organ* of benevolence. In short, in so far as they are organs they cannot be benevolent, and in so far as they are benevolent they cannot be organs.

As an embryonic science, phrenology did not long survive such criticism. Interest in the subject continued, and working phrenologists could be found throughout the century. J. W. Redfield of New York, for example, published in 1894 a list of 160 located faculties, among which there could even be found a faculty of republicanism. But by this time new techniques for the study of brain anatomy and function had been developed, and they pointed in quite different directions. In 1870 Hitzig and Fritsch reported the effect of electrically stimulating the cortex of a dog. They identified five areas which produced distinct bodily movements. Further work by David Ferrier, fully described in his *The Functions of the Brain* (1876), demonstrated the presence of numerous areas in the brain controlling, not *faculties*, but specific sensory and motor *functions*. After this work there remained no role for phrenology in science.

Parascience

In our own day the stranger seeking admission from outside the gates of science is the study of psychic phenomena, once known as extra-sensory perception, but becoming increasingly known as parapsychology, paraphysics, or, the term used here, parascience.[21] Partial though contested success was gained in 1969 when, largely at the behest of Margaret Mead, the American Association for the Advancement of Science granted membership to the Parapsychological Association.[22]

The serious study of psychic phenomena began in the mid-nineteenth century, after inexplicable rappings were heard in the house of a Methodist farmer, J. D. Fox, in Hydesville, New York. The rappings appeared to be related to the presence of his young daughters, Margaretta and Catherine; such phenomena were exported to Britain in 1852 by the New Jersey medium, Mrs W. R. Hayden.[23]

The opinions of scientists were soon sought. In 1853 Michael Faraday was invited to examine the claims of mediums and reported that table movements were very likely produced by the unconscious exertions of spectators' hands. Faraday's interests, however, were never more than casual. The leading British scientist to conduct a long and careful series of experiments into psychic matters was Sir William Crookes (1832–1919). An experimental chemist, the discoverer in 1861 of the element thallium, and a future president of the Royal Society, Crookes began as a moderate sceptic, confident that normal experimental controls would soon expose any trickery among those who claimed the power to levitate or to raise 50lb weights by acts of will alone.

Once established, the link with science has persisted. Of the forty-seven presidents of the Society of Psychical Research between its foundation in 1882 and 1982, some twenty-one have been scientists. Of these ten have been fellows of the Royal Society, and two have been Nobel laureates. The authority brought to any discipline by figures of such eminence has impressed many. Their honesty, competence, and technical insight are surely unquestionable. What possible grounds could anyone have for doubting the eye-witness report of Crookes dating from 1874:

The most striking cases of levitation which I have witnessed have been with Mr Home. On three separate occasions have I seen him raised completely from the floor of the room. Once sitting in an easy chair, once kneeling on his chair, and once standing up. On each occasion I had the full opportunity of watching the occurrence as it was taking place.[24]

It should be added that Home apparently demonstrated his powers of levitation in good light, and in Crookes's own home. Can anyone today have any plausible grounds for disbelieving Crookes on an issue as simple as this?

The point is one repeatedly raised by defenders of parascience. Hume had complained in his essay *Of Miracles* that supposedly supernatural and miraculous events were observed chiefly 'among ignorant and barbarous nations' and never by 'a sufficient number of men, of such unquestioned good-sense, education, and learning, as to secure us against all delusion in themselves'.[25] Hume's argument is no longer tenable. The literature of parascience is rich in eye-witness reports from men of 'unquestioned good-sense, education, and learning' of events as apparently miraculous as any considered by Hume. Whether these reports need be accepted unconditionally is in part a philosophical issue. Why do we place credit on witnesses and historians?, Hume had asked. How do we respond to a great authority who reports an incredible fact? Does the incredibility, Hume pondered, invalidate the report? Precisely these questions arise over the reports by Crookes and more recent figures.

There is no *a priori* connection between testimony and reality, Hume had answered. The reason we accept testimony is that 'we are accustomed to find a conformity' between it and reality. When, however, we face testimony of the incredible, even if it is presented by a great authority, we can no longer rely upon any such conformity.

There are, in fact, three issues to be faced in examples of this kind, each of them linked with one aspect of the viability of parascience. The first issue is concerned with the general question of the credibility of witnesses. Can we dismiss such evidence out of hand, or must we accept, in some cases at least, evidence of this kind? Secondly, there is the question of the methodology of parascience and whether or not it is significantly different from the methodology of science or other rational endeavours. And there is finally the matter of the implausible. Are psychic phenomena all that implausible? And, in any case, is this sufficient reason to dismiss them out of hand? The three points will be discussed separately.

Credibility of witnesses

The argument is frequently put by parascientists that someone of the stature of Crookes was known to be a careful and competent experimentalist. We can only, at this distance, reject his evidence in a completely arbitrary and dogmatic fashion. If the experiments of Crookes are to be rejected in this offhand manner then we would be equally justified in rejecting any other experimental report. Why, for example, do we feel free to accept without question the report of the astronomer Johann Galle that on the night of 23 September 1846 he sighted the planet Neptune for the first time, while feeling equally free to dismiss out of hand the report of Crookes that in good light, and in his own home, he had seen Daniel Home levitate?

Two points can be made in reply, one general and one particular. Men lie in all ages, Hume noted; and, he could have added, scientists and even fellows of the Royal Society are unlikely to be exempt from this particular human failing. Parascience itself has thrown up its fair share of frauds. S. G. Soal, President of the Society for Psychical Research in 1950, and Dr J. Levy, a Director of the Parapsychology Research Centre at Duke University, are just two leading workers who doctored their experimental results to yield evidence of psychic forces.[26]

More significant has been the case of the leading psychologist Cyril Burt (1883–1971). For twenty years the official psychologist of the London County Council, Professor of Psychology at University College, London (1932–50), and editor of the British Journal of

Statistical Psychology, Burt was for more than a generation one of the most powerful and respected modern psychologists. Much of the evidence for the view that intelligence was predominantly inherited had been collected by Burt. The crucial evidence came from comparisons of the IQs of separated identical twins. By 1966 he had apparently collected details of fifty-three such pairs. The correlation between their IQs, despite considerable variation in their upbringing, was claimed by Burt to be as high as 0.771. Burt's work was seized on by a number of talented but dogmatic disciples. All criticism of the hereditarian view was dismissed by them as uninformed, wishful thinking and, above all, as lacking the precision and rigour of the results collected by Burt.

Anomalies in Burt's results were first publicly pointed out by Leon Kamin of Princeton University in 1972.[27] Further investigation by Kamin in 1974 and by Oliver Gillie established that much of Burt's data was fabricated. The charges were confirmed by Hearnshaw in his sympathetic biography of Burt.[28]

There are, no doubt, many morals to be drawn from the Burt affair. The only one I wish to emphasise at this point is that unsupported human testimony, whoever it comes from, can always be challenged. Presumably it was the eminence of Burt which allowed him to publish results without providing 'even the most elementary information about how, where or when his purported data had been collected'.[29] Consequently, while it would be absurd to dismiss such testimony as worthless, it can do little more than draw attention to further problems and possible solutions. Testimony of this kind is, thus, simply evidence for the existence of certain phenomena; it is not, and never can be, conclusive proof of its reality.

What, secondly, of Crookes's particular qualities as a witness of psychic phenomena? Some critics have found them suspect. In 1874 Crookes began to investigate the medium Florence Cook. She often worked through the materialised form of Katie King, supposedly the daughter of the seventeenth-century pirate Sir Henry Morgan. During seances held in his home Crookes talked to Katie, photographed her, timed her pulse, walked arm in arm with her, and described her appearance and behaviour in the same factual manner in which he discussed the levitation of Home.[30] Katie's physical presence and Home's levitation were both equally real to Crookes.

Initially we were asked to accept the authority of Crookes's observations of levitation because he was a much-respected experimental chemist and an experienced observer. The same argument would apply also to the stroll he took, arm in arm, with a seventeenth-century incarnation. If Crookes could allow himself to be duped by Florence Cook then it seems more plausible to suppose, eye-witness reports notwithstanding, that he had also been duped by Donald Home. It can, of course, be objected that few observers would be exempt from charges of this kind. Similar oddities have been located in the lives of Darwin and Newton. Are their reports, therefore, not to be trusted? The point can be accepted without difficulty, for it merely reinforces the original claim that no unsupported testimony is beyond challenge.

Eminence is no defence. Critical examination of a number of Newton's own observations, for example, show conclusively that at certain crucial points he fudged his data. In the first edition of *Principia* (1687), Book 2, Prop. L, Newton calculated the velocity of sound to be 968 feet per second. This figure agreed well with measurements he had made in the cloisters of Trinity College. More careful measurements made by William Derham in 1704 yielded a velocity of 1,142 feet per second. Newton responded to Derham's work by introducing a number of extra features into his calculations which allowed him blandly to announce that 'sound will pass through 1142 feet in one second of time.' He added, as if to confirm his work, that experiments had shown that sounds 'do really advance' with this velocity.[31]

Methodology and parascience

What, then, of the experimental evidence offered by parascientists for the existence of psychic phenomena? It is conceded that many earlier experiments, judged by the standards of today, are worthless. It is also conceded that a fair share of today's work will turn out to be, for a variety of reasons, equally worthless. So much is also true of orthodox science. There remains, however, a bulk of work in both areas indistinguishable in quality and validity. Parascientists have adopted standards of evidence no less rigorous than those of orthodox science, use automatic recording instruments wherever possible, and subject their results to the strictest mathematical analysis. A typical research report will therefore read: 'In 794 calls there were 236 hits whereas the mean chance expectation was 159.4 ... The odds against this excess ... if there were no cause operating to produce right guesses in a series of this length is about 1000000 to 1.'[32] It is then pointed out that if comparable results had been achieved in orthodox science publication would automatically follow. Why, then does the parascientist alone find his work dismissed unread, unpublished, and ignored? To escape the charge of crass prejudice it is incumbent upon the critic to show how the methodology of parascience differs from that adopted by orthodox scientists.

The basic error of the parascientist is to misunderstand completely the nature of experimental science. He assumes that if he can only make his experimental procedures tight enough, if all loopholes are foreseen and avoided, then men of goodwill will respond to his work. The assumption is that there can be a *single* experiment so pure as to be beyond reproach. The critic however is more worried about the singularity of the work than its purity. Attempts to gain ever higher levels of purity lead, to the bafflement of the parascientists, to no discernible change in the attitude of the critic.

The critic, for example, is often referred to a paper by the physicist Helmut Schmidt.[33] His subjects were presented with four lamps and were asked to indicate, by pressing the appropriate button, which lamp would light up next. As the switch was controlled by a random quantum process no subject could be expected to score better than the chance expectation of one in four hits. In 63,066 trials 16,458 hits were scored, 691.5 above chance, with odds against of 10^8:1.

Let us assume that analysis reveals no major fault in the design and execution of Schmidt's experiment; does this mean that psychokinesis has been demonstrated? No, for it is a single experiment and, as such, is only the beginning of the process of demonstration and not its completion. It is casually assumed to be technically adequate, but what does such an assumption mean? Surely, in part, that the experiment in other hands can be replicated under a wide variety of conditions. The parascientist may insist that if the critic cannot point to some specific piece of fraud, carelessness, or faulty design in the experiment, he can have no grounds for rejecting it. But experience teaches us that experts, particularly when examining new equipment, often declare something free from defect moments before it completely disintegrates. Before the expert can guarantee that something is free from all likely defects he must have some idea of the full range of possible defects, and this is knowledge which can only come from experience. Scientists expect experiments to go wrong; that it will take time, effort, and the labour of many hands to get equipment and materials working properly and uninfluenced by extraneous factors.[34] To be presented with a single experiment and told it means that something quite specific has been demonstrated fits ill with their normal procedures.

Failure to appreciate this point is largely the fault of the textbooks of science. Here experiments are presented in neat little diagrams, proving or disproving some point of view. Little is ever said of the struggle to achieve the neat little diagram.[35] One of the most famous experiments in the history of science is the demonstration by Newton that when a ray of light is passed through a prism it reveals the colours of the spectrum spread out in 'an oblong form'. The experiment is now so familiar, the province even of young school-children, that it comes as a shock to realise that it took the finest scientists of Europe over forty years to appreciate, not merely its significance, but that Newton had really described what happened. Newton published the experiment in 1672. Several tried to repeat his work and failed. They thought that, perhaps, it was not so much the prism which was crucial but the presence of a nearby cloud through which the sun shone. Newton insisted that the experiment worked whether clouds were or were not present. His critics, however, insisted that the oblong form was 'never seen in a clear day, as Experience shews'. Other workers could obtain an oblong form, but one with different dimensions. The dispute continued until 1677 when Newton, his critics unsatisfied, simply refused to answer any further queries. Even a public demonstration before the Royal Society in 1676, and the publication of the *Opticks* in 1704, failed to convince his by then mainly continental critics. The acceptance of the reality of Newton's experimental work came only after a visit in 1714 of a number of Paris scientists to London. Newton arranged for his work to be demonstrated before them and, convinced at last, they returned home to demonstrate in turn his experiments to their colleagues.[36]

No doubt science today moves more rapidly than it did in the seventeenth century, but a number of features present in the example are still relevant. First, it is not immediately apparent what is and what is not relevant to the experiment. Do clouds matter? Unconsidered initially by Newton, their irrelevance had to be shown, and this could be done only by further experimental work. Secondly, disputes about the validity of the experiment could proceed quite independently of any theoretical issues. The great theoretical issue of the day was whether light was composed of waves or particles; the issue, however, seldom intruded into the dispute. Thirdly, whatever the arguments, the replication of the experiment by experienced scientists appeared to be vital. No doubt questions were put to Newton in 1714 by his French critics. It remains quite clear none the less that nothing is as convincing as a clear, public, and oft-repeated demonstration.

The obsession with replication is manifest in the contemporary and expensive world of high energy physics. This is a world in which claims for the existence of such elusive and exotic particles as the quark, the magnetic monopole, the Higgs boson, and a decayed proton are regularly made. The response of the profession is invariably the same; namely, a wary interest. This attitude is clearly shown in Close, Marten and Sutton's account of the subject. Of the claim by Fairbank to have detected a free quark, they note that no one else has reproduced a similar effect, and comment on the difficulty of eliminating *all* sources of background interference, finally noting that many physicists believe 'that some unknown source of error is responsible'.[37] On the claim by Blas Cabrera in 1982 to have discovered the magnetic monopole, they record only one further sighting, and conclude: 'Was this the real thing or some experimental effect that no one has yet been able to explain? Only time will tell.'[38]

What of the proton? According to theory it should have a lifetime of 10^{32} years. This means that if you can isolate 1,000 tons of the right kind of matter you should expect to see about 100 protons decay each year. To detect this Indian physicists at Bangalore, Japanese physicists at Kamioka, along with their French and American colleagues at Frejus and Mont Blanc, Minnesota, and Ohio, are all camped in deep mines scrutinising large swimming pools with numerous expensive detectors. So far the proton has remained stubbornly durable. What does this mean? Nothing other than that, the authors conclude, the possibility of proton decay remains open and will 'continue to test the ingenuity of physicists for some years to come'.[39]

Further insight into the significance of replication comes from Richard Feynman's candid memoirs of his career as a theoretical physicist. Describing what he terms 'poor science', he reports how a psychology student wished to work on the behaviour of rats. It was known that under certain circumstances rats did X. If the circumstances were changed, the student asked, would they still do X? First, Feynman insisted, it must be determined whether rats really did behave as described. But the student was told by his supervisor that it was 'general policy then not to repeat psychological experiments, but only to change the conditions and see what happened'.[40]

Similar practices, Feynman notes with 'shock', are creeping into physics. A colleague wished to see if results established for heavy hydrogen also held for light hydrogen. The National Accelerator Laboratory, however, refused to allow the earlier experiments with heavy hydrogen to be repeated. So anxious appear to be the administrators of NAL to achieve new results, Feynman laments, that 'they are destroying – possibly – the value of the experiments themselves', and diminishing the 'integrity' of experimental science. Feynman keeps his greatest scorn for the notion that there could be results of experimental science which required no replication: 'This is science?'.[41]

Nor is there any reason why parascience should be exempt from such demands. Consequently, before Schmidt's work can be accepted, it must be repeated. This is not simply to verify Schmidt's honesty but to see if factors unknown to him, or considered irrelevant by him, may in some other worker's hands prove quite vital. Attempts to replicate his work were made by John Beloff and D. Bate in 1971. Five subjects underwent 18,650 trials; the results were unsuccessful. This in itself is neither surprising, nor particularly significant. It is, in fact, the experimental norm to find some initial difficulties in replicating another's results. More surprising was the reaction of Beloff and Bate. The obvious conclusion was that only further experimental work could determine who was at fault. Beloff and Bate, however, were for no apparent reason convinced that their own failure 'to get positive results in no way detracts from Dr Schmidt's success'. Rather, they felt, that lacking 'Dr Schmidt's personal magnetism, his manner, authority', they had failed where he had succeeded in inspiring 'his subjects to greater efforts'.[42]

Their response is typical of parascience, and quite uncharacteristic of science. Presented with positive results, parascientists proclaim them to be evidence for the existence of psychic phenomena; failure to reproduce the results, however, is too frequently seen as lack of 'personal magnetism' on the part of the experimenters, or a result of the tenuous and delicate nature of psychic phenomena. They seem to be so tenuous and delicate that in fact the very attempt at replication causes them to dissipate completely. The physicist John Hasted claims that 'If the experimenter makes it obvious that he is scrutinising or is suspicious of the subject, then the effects are likely to be weak or altogether absent.'[43] At one place Hasted claims that psychic phenomena can be validated by 'orthodox techniques of physics'.[44] Elsewhere, he warns that in front of sceptics 'the atmosphere polarises, and then with the best will in the world, you get nothing.'[45] Consequently, in order to encourage psychics to work and produce results experimenters are often prepared to tolerate the most surprising concessions. Thus Hasted, wishing to test the power of a subject to scrunch up paper clips psychically, placed the clips in a glass globe. He found the subject 'could do this only when the globe contained a small orifice'.[46] Or, again, Soal and Bowden reported in 1959 on results obtained from two young Welsh schoolboys. When the boys could see some part of each other, significant results were obtained. When, however, the boys were separated by a door, they operated only at the chance level. The boys, the parascientists noted, were clearly 'suffering from a psychological inhibition with regard to closed doors'.[47]

With excuses and attitudes of this kind, it is not surprising that parascientists do not place the same emphasis on replication found among more orthodox scientists. The alternative is to create a single experiment so pure, so well monitored, and so automated, as to be instantly convincing. Such however is not the way of science. Built into the scientific consciousness is the deep realisation of how misleading, how false, how totally bogus, apparently successful experiments can be. Nor have the sophisticated instruments of today's laboratories improved matters.

So much is clear from the recent and highly instructive case of polywater. In the early 1960s a leading Russian physical chemist, Boris Derjaguin, described a strange form of water, found when water condensed in fine capillary tubes. It had a high viscosity and boiling point. The liquid was subjected to the most sophisticated spectral analysis then available in some of the world's leading laboratories, and was reported to be water, but so different from ordinary water as to be called 'anomalous water' and, by those who suspected they were seeing a totally undreamt-of polymer of water, 'polywater'. Within a few years, however, anomalous water had been shown to be no more than contaminated water, and the subject rapidly disappeared from the literature.[48] If a whole community of specialists working on something whose chemistry is well known, using the full range of modern analytic equipment, can be so disastrously wrong, what trust can be placed in a single experiment performed by Schmidt in 1969. The error of polywater was soon detected. This was possible because scores of chemists throughout the world were desperately trying to replicate the results of their colleagues. Eventually errors, ambiguities, misreadings, malfunctions, and a whole range of fairly typical mistakes were picked up and the fictitious nature of polywater was established. Unable to resort to such activities, the parascientist remains locked in his own experimental set-up. Such is undoubtedly not the way of orthodox science. Consequently, on methodological grounds alone, and excluding considerations of fraud and incompetence, the scientific pretensions of parascience can be confidently dismissed.

The plausibility of psychic phenomena

Psychic phenomena are inherently implausible. In their defence two points are commonly made. First, critics are advised against too hastily dismissing certain phenomena as fictitious. Thus, Lord Rayleigh, a President of the Royal Society and a Nobel laureate, in his 1919 presidential address to the Society for Psychical Research, cautioned 'those . . . who are so sure that they understand the character of Nature's operations as to feel justified in rejecting without examination reports of occurrences which seem to conflict with ordinary experience'.[49] Had not scientists in the past been scornful of those who had claimed that stones sometimes fell from the sky, and that the tides are controlled by the moon? It is certainly true that the chemist Antoine Lavoisier had reported in 1772 to the Académie des sciences that stones claimed by witnesses to have fallen from the sky were merely ordinary terrestrial stones which had been struck by lightning. It is also true that Galileo in his *Dialogue on the Great World Systems* (1632) had objected to Kepler's proposal that the tides were caused by lunar attraction, on the ground that Kepler 'had lent his ear ... to occult properties and such-like fancies'.[50]

The argument as it stands proves too much. That scientists have sometimes been premature and over-confident in dismissing new ideas is surely no reason why they should not be free to reject some theories without examination. For one thing the opposite view would make life very difficult. The libraries of today are rich in their collections of accounts of paranormal events, and to examine only a fraction would require the full-time commitment of many scientists. Pyramid power, divination by entrails, palm-reading and much more are argued for as seriously as any other paranormal event. Scientists must rank claims in terms of plausibility and relevance in order to decide which to investigate, and it is naïve to suppose any other arrangement could be made to work. For example, the magazine *Prediction* for February 1983 carries an article on *Reincarnation* by Georges Zagarresco in which he informs us that we may have had as many as 1,200 previous lives. By making a few arbitrary assumptions, and using dates of birth, he provides tables enabling anyone to infer details of their previous lives. Ginger Rogers, born 16 July 1911, we learn, was once an irritable Swiss male nurse, and I was born before in South Australia in 1875, where I became an egoistic astronomer. Does such nonsense require critical examination? Do all 1,200 previous lives have to be examined in case one, at least, may prove significant?

The fact is, in science as in life, we regularly and frequently dismiss without examination reports which conflict with our ordinary experience. The phone bill for £100,000, the thermometer showing a reading of 0°C in the middle of a heat wave, and the petrol tank reading empty just after it has been filled, are the kind of reports we are often presented with and which many of us ignore as a matter of course. Sometimes we pay for our confidence and find ourselves stranded on the motorway without petrol; more often, however, our initial judgement was quite correct.

Further, if Rayleigh can quote the misjudgements of Galileo and Lavoisier, and no doubt many others, the historian can respond with an equally long list of scientists making unwarranted claims for the existence of various bodies, forces, rays, planets, animals, and fluids. In these cases the scientists who ignored the claims for the N-rays of Blondlot, Gurwitch's mitogenetic rays, Ungar's scotophobin, Barkla's J-rays, Derjaguin's polywater, Le Verrier's Vulcan, and the entelechies of Driesch will have missed very, very little.[51]

Scientific judgements often appear dogmatic to outsiders. The scientist may casually declare that something cannot have happened after having considered the barest of details. The situation arises because scientists tend to work with a few assumptions so basic that they will be allowed by all to discredit any contrary observations, experiments, deductions or whatever. The commutativity of multiplication is so fundamental that if I know that $27 \times 53 = 1431$, and if I am told that $53 \times 27 = 1341$ I know there is a mistake somewhere. I need to make no calculations to reach this conclusion.

So too in the empirical sciences. Find a violation of the conservation of energy, or constants taking on strange values, and immediately faults will be assumed. Claude Bernard, for example, the great French physiologist, described his response to the claim of a colleague that a patient, 'a woman in good health . . . had neither eaten or drunk anything for several years'. Despite his colleague's claim that all precautions had been taken, Bernard dealt with the argument brusquely. The slightest knowledge of science, he noted, would show deception, as the claim amounted to saying 'a candle can go on shining and burning for several years without growing any shorter'.[52] Also relevant here are the already quoted doubts of Einstein that there could be forces, whether psychic or not, which did not vary in some way with distance.[53]

Secondly, it is often argued by parascientists that, compared with some parts of modern science, their own work appears quite pedestrian. Levitation, metal-bending, and telepathy seem relatively normal when compared with some of the results of quantum theory. So accustomed are we by now to a diet of the impossible that we no longer have any excuses for shunning parascience. At which point there is much talk of particles travelling backwards in time, of other particles separated by vast distances simultaneously influencing each other, and of cats existing in some curiously undetermined state. Thus, among many others, Arthur Koestler has commented that the deeper the modern physicist 'intruded into the realms of the sub-atomic and super-galactic dimensions, the more intensely he was made aware of their paradoxical and common-sense defying struc-ture, and the more open-minded he became to the possibility of the seemingly impossible'.[54] Quoted in justification are the extravagant claims of Jeans and Eddington, without acknowledgement of Stebbing's critical work,[55] that the universe is more like a 'great thought' than a 'great machine'. The suggestion is conveyed that somehow quantum theory has prepared the way for parascience by first exposing us to patterns of thought long resisted by narrow-minded classical physicists.

Do in fact the strange theories of quantum physics make parasci-ence any more plausible? Not to many of us whose understanding of quantum mechanics is insufficient even to recognise that it deals with strange phenomena. If, therefore, it is necessary that I should first understand quantum mechanics before recognising the plausibility of parascience, I am likely to be denied forever such recognition. If the results of parascience cannot be translated into language I can understand then I can safely ignore it, along with *Finnegan's Wake*, the *I Ching*, differential geometry, and that increasingly large class of things, literary, artistic, and scientific, which I now know I will never understand in this life.

Koestler's argument is most strange. The normal procedure to explain something obscure is to reduce it to something much simpler and clearer. In this case, however, Koestler seeks to justify the problematic by equating it with the obscure. Perhaps the assumption is that while we might find it difficult to accept one impossible event, it would at least immunise us against further shock, and make it easier for us to handle a second impossibility. It would, however, be just as plausible to argue that incredulity is rather limited, and having exhausted our stocks on quantum theory, there is none available to spend upon parascience.

Providence and design

Can the facts and theories of science be so marshalled as to demonstrate the existence of God? Many, whether scientists or not, have seen signs of God's workmanship in the natural order. Lichtenberg, for example, was apparently astonished by the fact that 'cats have two holes cut in their coat at exactly the place where their eyes are.' In a similar fashion the equally perceptive Robert Boyle was amazed at the good fortune of lambs to be born 'in the spring coincidentally with the growth of the new grass'. Comparable examples of the divine foresight can be found in the traditions of most cultures, whatever their stage of development.

It is met with in antiquity in the writings of Cicero. In his *The Nature of the Gods*, for example, he put the following argument into the mouth of Balbus, the Stoic:

When you see a sundial or a water-clock, you see that it tells the time by design and not by chance. How then can you imagine that the universe as a whole is devoid of purpose and intelligence, when it embraces everything, including these artefacts themselves and their artificers?[56]

Nearly 2,000 years later William Paley chose to begin his *Natural Theology* (1802) with a similar argument. Although the clock by this time had gained wheels, springs, and regulators, it suggested to Paley, as it had suggested to Balbus, the need for a designer of both the instrument and the universe.

The argument so far presented calls for no more insight into the workings of nature than that certain complex, integrated structures are more likely to have arisen from design than from chance. Such an insight, if it be so, is quite independent of any scientific tradition. It was none the less sufficiently prominent in antiquity to attract the critical attention of a number of philosophers. The Epicurean philosopher of the first century BC, Lucretius, in his classic poem *The Nature of the Universe*, insisted that 'Nature is free' and operated 'without the aid of gods'. What, then, of the signs of divine workmanship so apparent to other minds? They were swamped, Lucretius insisted, by the 'imperfections' visible to all. Among these he listed a predominantly mountainous earth, wild beasts, bogs, seas, torrid heat, and perennial frost. Such features, together with the pestilence brought by the 'changing seasons', and the freedom given to 'untimely death to roam abroad', convinced Lucretius that 'the universe was certainly not created for us by a divine power.'[57]

Similar arguments were presented by Cicero. Divine providence, he argued, fitted ill with the actual nature of life. History has shown the ruin of many a good man and the success of more than a few villains. How could the gods, he asked, allow the Stoic, Anaxarchus, 'to be pounded to death in a mortar by order of the tyrant of Cyprus', while permitting Harpalus, 'one of the most successful robbers of his time', to prosper?[58]

Newton

The objections of Lucretius and Cicero had little impact on western thought. Throughout the ages generation after generation continued to be impressed by the harmony of existence, and to see in such harmony the hand of God. In the seventeenth century, however, a new and more recognisably scientific dimension was added to the argument. It is seen very clearly in the work of Newton, and within that corpus it is presented most explicitly in his correspondence with Richard Bentley.

In 1692 Bentley had been invited to deliver the first of the newly founded Boyle lectures.[59] Set up by Robert Boyle, they charged the lecturer to prove the validity of the Christian religion against 'Infidels, Atheists, pagans, Jews, and Mahometans'. The foremost classical scholar of his day, Bentley was no scientist. Consequently, he approached Newton, who had, a few years earlier in his *Principia* (1687), deduced 'the motions of the planets, the comets, the moon, and the sea'. Could such a work, Bentley must have asked, ever serve to demonstrate the existence of God?

Newton replied that when he had written *Principia* he had in fact 'had an eye upon such principles as might work with Considering Men, for the Belief of a Deity'.[60] The principles Newton appealed to, however, were markedly different from those traditionally presented. In a famous line of verse written on Newton's death, Pope had claimed: 'God said, *Let Newton be!* and All was *Light*.' The line, in so far as it suggests that Newton offered to explain all aspects of the universe, is inaccurate. Indeed, Newton chose to stress, repeatedly, that it was the inability of his science to throw light on a number of issues that most clearly revealed the hand of God.

His argument on this matter was precise. Gravity and universal attraction, he emphasised, have limited explanatory power. There are areas, and physical areas at that, where appeal to gravity was pointless. The alternative could only be 'a divine hand'. He went on in his correspondence with Bentley to indicate in which areas God's presence was required. They include the following:

1 How could the original matter of the universe have been divided into two sorts, one of which, 'fit to compose a shining body, should fall down into one Mass and make a Sun', while the second sort should have composed not one but several opaque planets? It was inexplicable, Newton insisted, by 'meer natural causes' and could only be ascribed to 'the Counsel and Contrivance of a voluntary Agent'.[61]

2 Nor were there any natural causes capable of giving the planets 'those just Degrees of Velocity, in Proportion to their Distances from the Sun . . . which are required to make them move in such concentrick Orbs'.[62]

3 While gravity could account for the descent of the planets to the sun, their 'transverse motions . . . required the divine arm to impress them according to the Tangents of their Orbs'.[63]

4 'the Hypothesis of Matter's being at first evenly spread through the Heavens, is . . . inconsistent with the Hypothesis of innate Gravity, without a Supernatural Power to reconcile them, and therefore it infers a Deity.'[64]

5 The motions of the planets in the same plane and in the same direction could not be explained by any 'natural causes'; they must therefore be 'the Effect of Counsel'.[65]

Two things are immediately apparent. In the first place God's existence was being inferred from scientific facts and theories. The data presented went far beyond the run-of-the-mill illustrations quoted by earlier generations of scientists. While others had been content to see in the abundance of vegetables on the earth clear evidence of God's providence, Newton and his followers were more impressed by the close fit between a planet's velocity and its distance from the sun identified in Kepler's third law.

As a second point, Newton seems to have brought God closer to the operating mechanism of the universe than previous generations had contemplated. The earlier and more natural assumption had been that God, at the moment of creation, had laid down laws of nature sufficiently accurate and comprehensive as to allow the universe to continue harmoniously and of its own accord until judgement day. It was in this manner, for example, that Descartes had seen the role of God in the universe. The approach had been noted by Pascal, a contemporary of Descartes, and had led him to complain that 'In all his philosophy he would have been quite willing to dispense with God. But he had to make him give a fillip to set the world in motion; beyond this, he has no further need of God.'[66] The objection could not have been raised against Newton. He readily, indeed eagerly, accepted that the laws and definitions presented in *Principia* could not account for many important features of the universe. He further accepted that from time to time the universe could become so disorganised that it would need a divine 'reformation' to preserve its long-term stability.[67]

Nor was Newton the only figure of his day to see divine providence in nature. Other Boyle lecturers competed with each other to discover yet more features of the Newtonian cosmology and dynamics requiring divine intervention. Consequently, for a time such subjects as astro-theology and physico-theology, the titles of William Derham's 1711 and 1712 Boyle lectures, flourished as readily as more orthodox subjects.

Sources for the design argument other than Newton can be identified. The seventeenth century had seen the birth of a new mathematical discipline, probability theory. It too provided new arguments for God's providence. John Arbuthnott, yet another Newtonian, writing in the *Philosophical Transactions* for 1710, noted that the mortality bills for eighty-two consecutive years displayed in each case an excess of male over female births. The probability of such an outcome, Arbuthnott calculated, was $(\frac{1}{2})^{82}$. In other words, the odds against such a sequence happening by chance were quite astronomical. Arbuthnott went on to draw the inevitable conclusion. As males were normally subject to a greater mortality than females, 'To repair that loss, provident nature, by the disposal of the wise creator, brings forth more males than females, and that in almost constant proportion.'[68]

Before such an onslaught one contemporary theologian, Robert

Jenkin, noted in 1700 that 'infidelity could never be more inexcusable than in the present age.'[69] The reason for this should now be clear. Before the limits of the laws of motion could be established, before violations of the laws could be identified, the laws themselves had to be discovered and their implications worked out in some detail. This, however, was achieved only with Newton. Before Newton and Halley, for example, comets were thought to be transient, atmospheric phenomena, moving in an unpredictable path through the solar system. Once having disappeared from view they were gone forever. As their paths were unique they could threaten the security of princes while leaving the authority of the laws of motion unchallenged. But, once the orbits of comets and planets had been brought within the domain of scientific law, then and only then could the orbits of the heavenly bodies appear to be in any way exceptional.[70] In exactly the same way, only when mathematicians had worked out how to calculate the probabilities of particular events could it begin to occur to them that some actual events were wildly improbable.

Leibniz

Opposition quickly came from the philosopher Leibniz. It received its fullest expression in the *Leibniz–Clarke Correspondence* first published in 1717. Newton, Leibniz argued, had 'a very odd opinion concerning the work of God'. Lacking sufficient foresight to endow the solar system with perpetual motion, God seemed to be required to intervene occasionally in his creation and, as it were, 'wind up his watch from time to time'. As God's handiwork could not be imperfect, Leibniz insisted, any gaps in the system of *Principia* were the results of Newton's inadequate physics rather than God's oversight. The Newtonian response, presented through his disciple Samuel Clarke, rejected the notion of God as an absentee landlord, claiming that such an assumption led to materialism and banished God's providence from the world. The presence of God in nature, Clarke argued, was not 'a diminution, but the true glory of his workmanship, that nothing is done without his continual government and inspection'.[71] Thereafter both the Newtonians and the Leibnizians continued to argue with each other on the true implications of God's perfection, adding nothing, but simply repeating the positions of their masters.

Hume and 'the licence of conjecture'

A more comprehensively damaging and influential attack on the Newtonian position was presented by Hume in his posthumously published *Dialogues Concerning Natural Religion*.[72] Hume's consideration of the argument was detailed and definitive. The pivot on which much of the argument turned was his repeated insistence that 'Experience alone can point out . . . the true cause.' Thus when we see a house we can legitimately infer the existence of architects and builders because 'this is precisely the species of effect, which we have experienced to proceed from that species of cause.'[73]

If we wish to make a similar inference from the universe to its divine creator, we must resort to analogy. Chance can account neither for the assembly of bricks and mortar into a house, nor for the manner in which the planets move in the same plane and in the same direction. If in the first case we seek a builder, are we not equally entitled in the second case to see, in Newton's phrase, 'the Counsel and Contrivance of a voluntary Agent'?

Hume conceded the legitimacy of analogical reasoning, but went on to insist that while 'similarity of the cases gives us a perfect assurance of a similar event', departure from the similarity was liable to 'error and uncertainty'. For example, having noted the circulation of the blood in humans, we may confidently infer a similar system can be found in monkeys. That it also can be found in frogs and fishes is, however, 'only a presumption, though a strong one'. To conclude further, Hume warned, that sap circulates in vegetables, would be 'an imperfect analogy' and a mistake.[74]

How close, then, Hume asked, is the analogy between a house and the universe? It is so strikingly unlike, he declared, as to be no better than 'a guess, a conjecture, a presumption'. Arguments from experience work when we are in a position to note that 'two species of objects have always been observed to be conjoined together.' Consequently I can feel confident in supposing that the house before me must have had a builder. How, though, Hume went on to ask, can this argument be deployed when the 'the objects . . . are single, individual, without parallel'? To show that the solar system had risen from 'some thought and art', it would have been first necessary to have had experience of the origin of a number of solar systems.[75]

Once adopted on an inadequate basis analogy can lead anywhere. Just as plausible as the divine builder or clock-maker, Hume pointed out, were a number of analogies taken from the animal and vegetable world. The solar system could have arisen from an egg laid by a passing comet, or, following the Brahmins, we could maintain 'that the world arose from an infinite spider, who spun this whole complicated mass from his bowels'.[76]

The general view in which the universe was seen as comparable to a complex machine or a substantial house was dismissed by Hume as 'mere conjecture and hypothesis', or, in a phrase he used on a number of occasions, 'the licence of conjecture'. The phrase, I presume, was meant to contrast with arguments more soundly based on experience. The terms *conjecture* and *hypothesis* to an eighteenth-century mind suggested something quite different from the usage found in the pages of Popper and other contemporary philosophers of science. Newton, for example, had insisted that he did not feign hypotheses (hypotheses non fingo), and that 'Nature . . . affects not the pomp of superfluous causes.' Contrasted with talk of hypotheses, conjectures, and superfluous causes was the true method of science, by which 'we are to look upon propositions inferred by general induction from phenomena as accurately or very nearly true.'[77] Hume was, thus, using the language of Newton to overturn the general Newtonian position. As the conclusion of the design argument had not been inferred by general induction, and was not in fact based on experience, it could only be a hypothesis, a conjecture. As such it should not be confused with the mode of reasoning deployed in Newton's *Principia*, nor could it be supported by any of the scientific conclusions of that work. All were of course free to consider the origin of the universe. As soon, however, as they abandoned experience and resorted to such alternative approaches as analogical reasoning, experimental philosophy had been abandoned in favour of conjecture.

Properly seen, Hume's argument in the *Dialogues* was more general than its apparent subject-matter. Behind the design argument there lay the more general issue of the limits of science. His answer, that only propositions based on experience could be supported by science, meant that Newtonian physics could be made to count neither for nor against the existence of God. The argument itself remained largely unappreciated by scientists and theologians alike.

Kant

Hume's views were to be found echoed in Kant. In his *Critique of Pure Reason* (1781), he argued against the possibility of a 'Physico-Theological Proof' for the existence of God. In a manner reminiscent of Hume he objected that such a proof could at most 'demonstrate the existence of an architect of the world, whose efforts are limited by the capabilities of the material with which he works, but not a creator of the world, to whom all things are subject.'[78]

Paley and the Bridgewater Treatises

Many later writers, however, simply ignored Hume's arguments. Paley, for example, in his *Natural Theology* (1802), developed the design argument at some length and without reference to Hume's work. Unlike Newton, he was little impressed by astronomy.[79] Lacking parts and all complexity, heavenly bodies appeared as no more than luminous points. The world of biology was far more suggestive. Paley did more than list favourable examples, going on to identify a number of general mechanisms and structures. It was frequently noted that the defects of one organ were remedied by another. By a process of *compensation*, as he termed it, the disadvantages of the elephant's short neck could be overcome by a long and flexible trunk. By a second feature, *prospective contrivance*, the provision of such unnecessary organs as foetal lungs could be explained. By a third process, *relations*, Paley claimed to be able to show how such structures as male and female genitalia were, like lock and key, 'manifestly made for each other'.

The twenty editions of *Natural Theology* which appeared between 1802 and 1820 might suggest that the definitive work on the subject had appeared. The design argument, however, was pursued at even greater length in the Bridgewater Treatises of the 1830s. When the eccentric eighth Earl of Bridgewater died in 1829, he charged his executors to select eight authors capable of demonstrating 'the Power, Wisdom, and Goodness of God', as displayed in 'the variety and formation of God's creatures in the animal, mineral, and vegetable kingdoms'. To finance the project £8,000 was set aside, and by 1836 eight substantial volumes had been published. The emphasis throughout was on science. Sir Charles Bell (1774–1832), a leading neurologist, wrote on *The Hand* (1833) and how it evinced design; William Whewell (1794–1866) on *Astronomy and General Physics Considered with Reference to Natural Theology* (1833), and the chemist William Prout (1785–1850) on *Chemistry, Meteorology, and Digestion* (1834). Other scientific contributions came from Peter Roget (1779–1869), secretary of the Royal Society, the geologist William Buckland (1784–1856), and the physician John Kidd (1775–1851).

Two critics

Two Victorian scientists, however, did take Hume seriously and sought to refute his argument. The first, Charles Babbage, a pioneer of the computer, was one of the great scientific mavericks of his day. Dissatisfied with the Bridgewater Treatises, he issued, uninvited, a *Ninth Treatise* (1837) of his own. Hume had argued in his essay 'Of Miracles' that, as a miracle is 'a violation of a law of nature', then 'as a firm and unalterable experience has established these laws, the proof against a miracle, from the very nature of the fact, is as entire as any argument from experience can possibly be imagined.'[80] It is, thus, a miracle 'that a dead man should come to life', as such an event 'has never been observed in any age or country'.

Babbage, excited with the workings of his new 'Calculating Engine', deployed it against Hume's argument. He took the case of Lazarus, and asked how improbable it really was. Babbage calculated that with a human race 6,000 years old, and thirty years to a generation, we have had since Adam 200 generations. Allow the average population of the earth to be a billion, and we will find that since creation about 200 billion human beings have been born and have died. The odds against any one of them being restored to life is therefore 200 billion to 1.

Yet, Babbage insisted, he could programme his machine to

produce, under law, events just as unlikely. It would be possible to set the machine:

> so that it shall calculate *any algebraic law whatsoever*: and also possible so to arrange it, that at any periods, *however remote*, the first law shall be interrupted for one or more times, and be superseded by *any other law*; after which the original law shall again be produced, and no other deviation shall ever take place.[81]

For example, the machine could print 1, 2, 3, 4 . . . 200 billion, at which point, and for the next hundred places, the numbers increase by 2, before reverting forever to the initial rule that each number is greater by 1 than its predecessor.

'What would have been the chances against the appearance of the excepted case immediately before its occurrence?', Babbage asked. It would have had, according to Hume, 'the evidence of all experience against it'. The relevance of Babbage's argument remains unclear. By showing that the extraordinary could sometimes happen, Babbage had made it no more ordinary. Hume would have had little difficulty in rephrasing his argument to deal with Babbage's objection. An event which occurred once in 200 billion times could no longer be classed as something which 'has never been observed in any age or country'. It could, however, be seen as an event so rare and improbable that experience still constituted a proof against miracles.

Babbage, by putting himself in the position of the designer-programmer, has begged the question. To the observer reading the computer print-out, things would be less clear. With more experience of the ways of computers, a modern observer would simply assume yet another computer error. After all, when I receive an electricity bill for £100,000 I do not take this as evidence that the laws of thermodynamics have suddenly changed. I merely conclude, as before, that computers cannot be trusted.

A second and more forceful attack came from Hugh Miller (1802–56), journalist, stonemason, poet and geologist. Miller saw in the geological record, as did many of his contemporaries, evidence of a series of cosmic catastrophes, as the fossil record repeatedly revealed the sudden appearance and equally sudden disappearance of numerous extinct life forms. Hume had objected that arguments from experience could not be applied to singular events like the creation of the universe. In order to cover such an event we would first need to have experienced 'the origin of worlds'. It was insufficient merely to have seen 'ships and cities arise from human contrivance'. But, Miller replied, precisely such evidence is available in the testimony of the rocks. Geology breaks down 'that *singularity* of effect' on which Hume had built his case, and geology provided us not with one creation, but with many. And further, Miller concluded, 'It gives us exactly that which, as he truly argued, his contemporaries had not, – an *experience* in creations.'[82]

Let Hume apply his argument to each creation in succession, Miller demanded, and see where it leads. Initially there were no life forms on earth. The Humean would therefore have concluded, incorrectly, that 'It would be unphilosophic to deem Him adequate to the origination of a single plant or animal.' When the pines and reptiles appeared the Humean would still have insisted that 'It would be rash to hold, in the absence of proof, that He *could* originate aught higher and more perfect.'[83] And so, having made equally mistaken judgements over the appearance in the fossil record of the birds and the mammals, the Humean would find that man eventually appeared. An argument so misleading, Miller suggested, should be discarded.

The fallacy in Miller's argument consists of nothing more than confusing truth with validity. I cannot infer from a painting of Zeuxis alone, Hume had pointed out correctly, that he is also a sculptor or architect. If it is later revealed to me that Zeuxis is in fact a sculptor, then I have discovered a new truth about the world. The argument 'Zeuxis is a painter; therefore he is a sculptor' remains invalid, despite his skill with the chisel. If it were valid then it would follow that all painters are sculptors, a manifest falsehood.

What, though, of the sequence? If time after time I have underestimated the creator's power, if time after time a more advanced form has always appeared, would it not be reasonable this time to suppose that the creator had powers greater than those revealed in his creation? Hume, however, could well have replied that an unwarranted assumption is being made that a single creator is responsible for the reptiles, birds, mammals, and man. It is just as reasonable to assume that after having created the reptiles the creator had, like a distant pharaoh contemplating his completed pyramid, exhausted both his potentialities and his life. Why can we not suppose that different life forms are like the pyramids and constructed by different limited creators at different times?

Miller's *Testimony of the Rocks* had appeared in 1857. Shortly afterwards a more potent attack on the design argument as it was presented by Miller and in the Bridgewater Treatises appeared in Darwin's *Origin of Species* (1859). Darwin himself, as a student, had read and been impressed by Paley's arguments. Once having demonstrated the reality of evolution by a process of natural selection, it became apparent to him that such an approach was no longer plausible:

We can no longer argue that, for instance, the beautiful hinge of a bivalve shell must have been made by an intelligent being, like the hinge of a door by a man. There seems to be no more design in the variability of organic beings, and in the action of natural selection, than in the course which the wind blows.[84]

Darwin's alternative explanation of design proved to be more potent in removing the attractions of providentialism from scientists and theologians than the subtleties of the Humean position.

It is precisely this message that has been spelt out in considerable detail by Richard Dawkins in his *The Blind Watchmaker*.[85] Although Hume was right to criticise the manner in which scholars had taken 'apparent design in nature as *positive* evidence for the existence of God', he had failed to offer any 'alternative explanation' for this design. The gap has been filled by Darwin, with natural selection serving as the 'blind watchmaker', blind because 'it does not see ahead, does not plan consequences, has no purpose in view.'[86]

Modern physics and design

Free from the constraints of natural selection, physicists and cosmologists have continued to speculate about the order of nature. While biologists no longer feel the need to introduce divine providence into discussions on the evolution of the mammalian eye, physicists have begun to see in the detailed fine-tuning of nuclear and electromagnetic interactions clear evidence of design. Freeman Dyson, for example, one of the founders of quantum electrodynamics, has included in his autobiography a chapter entitled 'The Argument from Design'. Having first listed various numerical accidents that seem to conspire to make the universe habitable', he commented: 'I do not claim that the architecture of the universe proves the existence of God. I claim only that the architecture of the universe is consistent with the hypothesis that mind plays an essential role in its functioning.'[87] Consider one such accident, namely, the strength of the attractive nuclear force. It is apparently 'just sufficient' to overcome the electrical repulsion between the protons in the nuclei of such atoms as oxygen and iron. If the nuclear force had been 'slightly stronger', hydrogen would be rare and stars like the sun could not have formed; if, on the other hand, the nuclear force had been weaker, hydrogen would be incombustible and there would have been no heavy elements. But as, Dyson concluded, 'the evolution of life requires a star like the sun . . . then the strength of nuclear forces had to lie within a rather narrow range to make life possible'.

A similar argument has been deployed on numerous occasions by Fred Hoyle. One example concerns the production of the heavy elements. Astrophysicists had long sought processes which would lead to the production of the elements carbon and oxygen within the interiors of sufficiently hot stars. Carbon (C^{12}) could most naturally be formed by the fusion of three helium nuclei (He^4). This process, however, was too rare to account for the amount of carbon found in the universe. Hoyle and his colleagues struggled in the 1950s to work out the details of a more productive process. Suppose two helium nuclei ($He^4 + He^4$) fused to form a single beryllium nucleus (Be^8). This in turn could capture a further helium nucleus to form carbon ($Be^8 + He^4 \rightarrow C^{12}$). As a final step, C^{12} could capture another He^4 to form oxygen, O^{16}. The processes, as described above, seem straightforward and simple. Yet, for them to occur, Hoyle emphasised, it is necessary that 'Be^8 is unstable, that a resonant level exists in C^{12} at exactly the right place, and that a potentially dangerous resonance in O^{16} happens to be just below threshold.'[88] Without these 'three apparently more or less random accidents' neither the synthesis of complex nuclei, nor life, could have developed. Hoyle left his reader in no doubt what these facts meant to him: 'I do not believe that any scientist who examined the evidence would fail to draw the inference that the laws of nuclear physics have been deliberately designed with regard to the consequences they produce inside the stars.'[89]

Do we, though, have to accept the initial premiss of the argument, let alone the deduction? We are told that if the nuclear force had been slightly stronger, or beryllium more stable, then complex molecules could not have evolved. But, surely, if we can allow one feature of the system to vary, why not others? If the nuclear force had been stronger, could not the properties of hydrogen also have been sufficiently different to allow the gas to form readily under these new conditions? And if beryllium had been more stable, perhaps also the properties of helium could have been significantly different and complex nuclei could still have been formed.

A further attempt to identify design in the universe is based on the frequency with which certain key numbers seem to crop up in the most unlikely places. Consider, for example, the following three dimensionless numbers.

1   The age of the universe in nuclear units. This is obtained by dividing the age of the universe, 10^{10} years, by the time light takes to cross the proton, 10^{-24} seconds. The answer is 10^{40} years.
2   The force constant. What is the ratio between the gravitational attraction between two protons, and the electrical force between two charged particles? Again, the answer is 10^{40}.
3   The number of particles in the universe. Divide the total mass of the universe by the mass of the proton, and the answer is 10^{80}, or, more significantly, $(10^{40})^2$.

Many other numerical relationships and apparently benign coincidences have been identified by scientists, enough in fact to serve as the subject-matter of a whole book, Paul Davies's *The Accidental Universe* (Cambridge, 1982). Few were bold enough to infer directly from such puzzling data the existence of God. More, however, were convinced that the data were not without significance and, consequently, began to give expression to this insight in a variety of propositions now conveniently referred to as the Anthropic Principle.

The earliest statement, known as the Weak Anthropic Principle, is traceable back to Robert Dicke in 1957. The principle simply acknowledges that many of the features of the universe we observe, such as its size and age, must fall within certain limits. Otherwise we would not be here to observe them. Thus, the weak principle states:

The observed values of all physical and cosmological quantities are not equally probable but they take on values restricted by the requirement that there exist sites where carbon-based life can evolve and by the requirement that the Universe be old enough for it to have already done so.[90]

Given the principle, we can infer that the universe must be old enough to have allowed hydrogen and helium to be converted into the heavier elements essential for life within the interiors of the stars. But the production and dispersal of this material requires at least 10 billion years. Therefore, in the words of Barrow and Tippler, 'We should not be surprised to observe that the Universe is so large. No astronomer could exist in one that was significantly smaller. The Universe needs to be as big as it is in order to evolve just a single carbon-based life-form.'[91]

In this form the principle seems unexceptional. If it really does take 10 billion years to make the universe fit for carbon-based life forms, and as we are indeed forms of this kind, then, given the correctness of our assumptions, the universe must have reached this minimum age. In this form, of course, the principle has nothing to say about the presence or absence of design in the universe.

Some, however, have found the weak principle too limited. Brandon Carter in 1974 proposed the more speculative Strong Anthropic Principle (SAP): 'The universe must have those properties which allow life to develop within it at some stage in its history.'[92] This, however, even if true, would prove very little of interest. For example, it gives no indication of what the properties might be that allow life to develop. We can all make a few suggestions, and theoretical physicists may even be able to provide a fairly extensive list of properties, Pn, which must have been present in the big bang. In the same way, equally uninformatively, I propose the Strong Archosaur Principle: 'The universe must have those properties which allowed the destruction of the dinosaurs at some stage in its history.' Nothing follows from this about the nature of those properties.

The SAP does not even allow us to infer that the properties Pn must constitute a unique set. Given that a man is dead, we can infer that there must have been some state of the universe which led to his death. But, of course, any one of hundreds of different states, not all compatible, could have had this effect. Again, we seem to be left with the truism that if something happened, it did so because of something else.

Thirdly, and even more seriously, the SAP does not guarantee that even if the universe does have the relevant properties Pn, life must have developed at some stage. For it is often the case that the conditions for producing an outcome F are all present without F ever appearing. All the conditions necessary for the growth of my tulip may well be present. Good soil, plenty of fertiliser, adequate water, controlled temperature, and whatever else a tulip may need, are all provided in my carefully designed greenhouse. But, just before the tulip seed is about to germinate, an unexpected hurricane totally destroys the greenhouse and everything in it. In a similar manner the universe could well have had all the properties necessary for the creation of life without life ever developing. Perhaps, some 600 million years ago, just when life was about to appear on earth, an increase in the level of ultra-violet light, or a sudden burst, perhaps, of gamma rays from some distant nova, abruptly destroyed all incipiently vital forms.

The most surprising feature of the design argument, however, is its persistence. Despite radical changes in science, the scientific form of the argument is espoused by a number of physicists today as strongly as it was held by Newton some three hundred years ago. The examples have changed. Physicists are no longer disturbed by the supposed instability of the orbits of Jupiter and Saturn, and are now more likely to see signs of design either at the cosmic or the nuclear level. When, in the future, these events appear as natural as planetary orbits appear to the physicists of today, future scientists will no doubt have discovered some further feature of the universe in need of a designer. Perhaps, even, the day of the economist will have dawned, and he will see in the operations of the market the same evidence of design as physicists had previously noted in the heavens and in the atomic nucleus.

PART V

TRUTH, REALISM, AND
RATIONALITY

10

COULD SCIENCE HAVE DEVELOPED
DIFFERENTLY?

This monism or complete idealism invalidates all science. If we explain or judge a fact, we connect it with another; such linking, in Tlön, is a later state of the subject which cannot affect or illuminate the previous state. Every mental state is irreducible: the mere fact of naming it – i.e. of classifying it – implies a falsification. From which it can be deduced that there are no sciences on Tlön, not even reasoning.

Jorge Borges, *Tlön, Uqbar, Orbis Tertius* (1961)

The struggle in the West between religion and philosophy is well known and has been amply documented. Philosophy has its martyrs, heroes, and even the occasional villain. Among its martyrs is Giordano Bruno, burnt at the stake in Rome in 1600, while foremost of its heroes is Galileo, victim in 1633 of the Inquisition.[1] Yet Christianity was never a serious challenge to science. Most scientists, until relatively recent times, have been as devout as any cleric and have tended to see in nature not a challenge to, but a demonstration of, God's existence.

There were, however, other and more dangerous challenges to the authority of science. To a renaissance scholar, science came in a variety of forms. It was, further, just one out of several possible ways of looking at nature. Three traditions in particular proved especially challenging. Aristotelian thought still had sufficient vigour in the early seventeenth century to require scrutiny and critical analysis. Serving as a constraint on the development of modern science rather than a direct challenge to it, such concepts as substantial forms and final causes had to be convincingly discarded before science could flourish. A second challenge came from the magi who thought their own occult sciences to be more effective than the still developing natural sciences. And, finally, a revival of the sceptical tradition had led many to doubt whether it was possible to know anything.

Before the science of Newton and Boyle could establish itself, the pretensions of the Aristotelians, magi, and sceptics had first to be considered. While it may well be obvious to us that science had little to fear from its competitors, the picture must have looked very different to the scholars of the early seventeenth century. This battle also had its heroes. Foremost was Marin Mersenne (1588–1648), described by Popkin as one of the most important and neglected figures in the history of thought. Others involved were Gassendi, Gilbert, Descartes, Galileo, and, eventually, Locke and Newton. They struggled to demonstrate that an approach to nature based on observation, experiment, and calculation was preferable to any visible alternative. As science had yet to reach a state whereby it could justify its existence by appealing to its practical benefits and triumphs (synthesised steroids lay well into the future), claims for its superiority had to be based on other grounds. Inevitably those grounds turned out to be more philosophical than scientific, and centred on questions of method rather than content, and on what could and could not count as genuine cases of knowledge. The survival and development of science in the sixteenth and seventeenth centuries was, therefore, as much a part of the history of philosophy as of the history of science. We will begin with the occult.

The occult challenge

To a number of serious sixteenth-century scholars science proved to be far from attractive. Thus, Cornelius Agrippa (1486–1535), author of one of the most important works on magic to appear during the renaissance, also published a work on *De Vanitate Scientarum* (1530), in which all the sciences were dismissed as pointless. Astronomy, for example, was seen as concerned with 'vain disputes about Eccentricks, Concentricks, Epicycles, Retrogradations, Trepidations, accessus, recessus . . . the works neither of God nor Nature, but the Fiddle-Faddles and Trifles of Mathematicians'.[2] Similar thoughts had been expressed earlier by Erasmus in his *Praise of Folly* (1515). Scientists were accused of seeking to explain such inexplicable events as 'thunderbolts, winds, eclipses', and pursuing such pleasant forms of madness as 'measuring the sun, moon, stars and planets by rule of thumb'. Such folly, Erasmus pointed out, was revealed by 'the endless contention amongst [scientists] on every single point'.[3]

The nature of the alternative tradition can be readily seen in della Porta's *Natural Magick* (1589). He distinguished between two sorts of magic: the first is 'infamous' and deals with 'foul spirits', and is known as sorcery; the second is natural magic and is accepted by men of learning. Natural magic, Porta explained, worked 'by reason of the hidden and secret properties of things. There is in all kinds of creatures a certain compassion . . . which the Greeks call Sympathy and Antipathy . . . For some things are joined together as it were in a mutual league, and some other things are at variance and discord.'[4] Natural magic was, in fact, merely the practical part of natural philosophy, and with its aid 'we might gather many helps for the uses and necessities of men.'

Detailed rules were presented to enable the magus both to discern the hidden sympathies and antipathies, and to discover any practical use they might have. The key lay in 'the likeness of things', the famous doctrine of signatures by which 'the secret vertues' of things were revealed by their likenesses. Thus, Porta argued, 'the herb Scorpius resembleth the tail of the scorpion, and is good against the bitings', and coleworts, which shun the vine, are a cure for drunkenness. It was also the case that like attracted like, and the greater the smaller. Hence the heroic cure for bad breath proposed by Sir Kenelm Digby, a founder member of the Royal Society:

Everyone knows, for example, that a greater stench will attract a lesser. 'Tis an ordinary remedy, though a nasty one, that they who have ill breath, hold their mouths open at the mouth of a privy, as long as they can, and by the reiteration of this remedy, they find themselves cured at last, the greater stink of the privy drawing unto it and carrying away the lesser, which is that of the mouth.[5]

Digby is well known to students of seventeenth-century literature for his defence of the absurd sympathetic powder.[6]

The language of sympathy could also be used in more recognisably scientific ways. It could, for example, be used to explain magnetic attraction. Thus Porta argued that as the lodestone was a mixture of stone and iron, and as it contained more stone than iron:

Therefore the iron, that it may not be subdued by the stone, desires the force and company of iron; that being not able to resist alone, it may be able by more help to defend itself . . . The lodestone draws not stones, because it wants them not, for there is stone enough in the body of it.[7]

Other scholars were also prepared to defend their use of sympathies and other occult qualities. One such was Duarte Madeira Arrais, a physician from the University of Coimbra (in present-day Portugal), in his book *De Qualitatibus Occultis* which was published in 1650, the year of Descartes's death. The power of remora to slow ships, of eels to produce shocks, and magnets to attract iron, he argued, could not be explained in terms of the familiar four elements of earth, air, fire, and water. None of the elements attracted in these ways, nor did they produce their effects so rapidly, efficiently, and forcefully. Consequently, he concluded, certain bodies did possess occult powers.[8]

Also available was a comprehensive theory of magic, usually in neo-platonic terms, and showing its connections with more orthodox divisions of knowledge. Thus, Agrippa began his *De Occulta Philosophia* (1533) with a division of the universe into the intellectual, celestial, and elemental worlds.[9] Power flowed downwards from the angels of the intellectual world, via the stars of the heavens, to the matter of the elemental world. Corresponding to each was a different kind of magic. The system and its interconnections can be clearly seen:

Worlds	Type of magic	Medium	Related discipline
Intellectual	Ceremonial or ritual	Angels	Theology
Celestial	Astrology	Stars	Mathematics
Elemental	Natural magic	Elements	Physics

Such a scheme obviously opens up a role for numerology. It was quickly adopted by magi, mathematicians, and scientists alike. Copernicus had reduced the number of planets from seven to six by showing that the moon was simply a satellite of the earth. Rheticus, a disciple of Copernicus, immediately saw in this a numerological basis. The number 6 was the smallest perfect number; it must therefore have been pleasing to God, he argued, 'that his first and most perfect work should be summed up in the first and most perfect number.'[10]

Against such modes of reasoning a number of distinctions had to be made, and a number of objections had repeatedly to be raised. First, the basis of many of the supposed sympathies had to be challenged. Thus, Mersenne opposed in a quite natural way the traditional identification of the seven planets and metals: Saturn with lead, Jupiter with tin, Mars with iron, copper with Venus, quicksilver with Mercury, and silver and gold with the moon and the sun respectively. The identification was based upon the colours, weights, and motions of the various planets and metals. Thus, Saturn, the slowest planet, is obviously leaden, and Mercury, the quickest, moves like quicksilver. Yet, Mersenne asked, are the correspondences genuine? Lead is not the heaviest metal; gold and quicksilver are heavier. Should not Saturn, therefore, be golden? As for the leaden colour of Saturn, Mersenne argued, it could equally be called silvery. And, in any case, the colours of the planets were caused, not entirely by the planets themselves, but partly by the various media through which the planets' colours are discerned. The correspondences are, in short, arbitrary.

A further principle, that like attracts like, was similarly challenged by Gilbert in his *De Magnete* (1600). Likeness could not be the cause of amber's attractive powers, Gilbert pointed out, for it attracts 'all things that we see on the globe, whether similar or dissimilar'. More generally, 'Besides, like does not attract like – a stone does not attract a stone, flesh flesh: there is no attraction outside of the class of magnetic and electric bodies.'[11]

Secondly, the explanatory language of natural magic was dismissed as merely metaphorical. Thus Gilbert objected to Porta's account of the lodestone quoted above: 'It is the height of absurdity to speak of these substances . . . as warring with each other and quarrelling, and calling out from the battle for forces to come to their aid.'[12] Talk of 'fights, seditions, conspiracies in a stone' could be no more than 'the maunderings of a babbling hag'. What of sympathy? Gilbert was equally dismissive: 'But even were fellow-feeling there, even so, fellow-feeling is not a cause; for no passion can rightly be said to be an efficient cause.'[13]

Finally, and most important of all, there came the growing realisation that natural magic was concerned with no more than symbolic or verbal connections. Orchids are shaped like testicles, and kidneywort like kidneys. Will orchid juice, therefore, as Paracelsus maintained, 'restitute his lewdness to a man', a kidneywort cure renal failure? Or have kidneys and kidneywort no more in common than a name? The motto of the Royal Society would be 'Nullus in verba', or 'Nothing in Words', an insight recorded by many earlier writers. Thus Kepler took Fludd to task, reminding him that 'Nothing is proved by symbols; things already known are merely fitted [to them]; unless by sure reasons it can be demonstrated that they are not merely symbolic but are descriptions of the ways in which the two things are connected.'[14] Bacon also, in *The Advancement of Learning* (1605), listed as 'the first distemper of learning, when men study words and not matter'.[15]

The occult tradition was immensely helped by the prevailing Aristotelian account of causation. The main advantage to the magi lay in the sheer multiplicity of possible causes. In the production of any object or event, Aristotle argued, we can recognise four causes. We can, for example, ask of a table 'What is it made from?' and thus seek to identify the *material* cause. Or we can ask 'What is its shape or structure?', and so look for its *formal* cause. Again, the question 'Who made it?' will lead us to its *efficient* cause. And, finally, the question 'What is its purpose?' will involve us in considerations about the table's final cause. The theory worked reasonably well when applied to the construction of such artifacts as tables and chairs, and could even be used to explain natural processes like the growth of plants. How, though, can we explain in these terms magnetic attraction, the tides, and the power of kidneywort to cure renal disease? Before progress could be made on these and other issues, Aristotle's theory of causation, along with a number of other survivals from the Aristotelian canon, had to be critically scrutinised.

Aristotelian residues

The would-be scientist of the renaissance found himself working with a number of concepts and distinctions which, though deriving ultimately from Aristotle, had been reshaped by generations of scholastic philosophers. Some of these concepts proved to be obstacles to the emergence and development of science and required careful philosophical excision before science could begin to flourish. Four theories in particular seemed central to any understanding of the natural world – multiplication of species, substantial forms, the four elements, and the four causes – and their impact on science will be examined briefly.

Multiplication of species

Early atomic theories of vision proposed that something material – known variously as an *eidolon* or *simulacrum* – was ejected by the visible object and transmitted to the eye of the viewer. The theory soon had to face some massive objections. How, it was asked, could so many material replica fill the air without constant interference? Or, how could the simulacrum of a large and only too visible mountain shrink sufficiently to enter the eye of the beholder? Against such evidently compelling objections Aristotle argued that the sense organs receive the form, but not the matter, of the perceived object. Distant objects impress their forms on the medium, as a signet ring imposes its image on wax, which in turn are transmitted by the medium to some distant eye.

The model proved too attractive to be restricted to vision and the other sense organs. Roger Bacon, for example, in the thirteenth century, argued accordingly that species (outward appearance, form, or visual image) could also be deployed to explain the transmission of heat and sound, magnetic attraction, stellar influences, and, indeed, any other instance where things seemed to act distantly.

The disadvantages of the approach are apparent. By providing an explanatory form capable of handling virtually all interactions, differences are ignored, problems obscured, and very little actually explained. 'A natural agent multiplies its power from itself to its recipient, whether it acts upon sense or matter', Bishop Grosseteste declared in the thirteenth century in his *De Lineis*, a principle broad enough to cover virtually any event, anywhere in the universe.[16] Although essentially naturalistic, species as an explanatory mode seems no improvement on the claim that agents act upon distant objects through the will of God. Just as any natural event can be assigned to God's will, so too can they, equally unthinkingly, and just as automatically, be accepted as yet another case of the multiplication of species.

But before any kind of progress could be made with some of the central problems of science, they first had to be seen not only as problems, but as difficult problems. If magnetism, vision, heat, sound, etc., could all be explained by referring to the transmission of magnetic and other kinds of species, the problems of science would not even be recognised. One important and preliminary task facing philosophers, if science was ever to thrive, was a careful consideration of the nature of species.

Thus, Descartes in his *Opticks* (1637) began by arguing that in vision it was quite unnecessary to suppose that something passes between object and eye, or even that there is anything in the object which resembles the ideas and sensations we have of them. For example, when a blind man feels something with his stick:

> Nothing has to issue from the bodies and pass along the stick to his hand; and the resistance or movement of the bodies . . . is nothing like the ideas he forms of them. By this means, your mind will be delivered from all those little images flitting through the air, called 'intentional forms', which so exercise the imagination of the philosophers.[17]

Leibniz, also, in his *Monadology* (1714), found the notion unhelpful. If we are to suppose that we recognise an object as green by receiving a green species or form cast off by the object, then we are committing a logical blunder. Accidents, Leibniz argued, cannot exist separately, apart from substances. They cannot 'become detached, or wander about outside a substance'.[18] We could not, that is, perceive something green without perceiving a green something.

Substantial forms

Linked to the notion of species was the traditional doctrine of substantial forms. Any individual, for Aristotle, was composed of matter and form. In the simplest case, gold matter could be shaped into the form of a ring, and marble into a bust of Socrates. But things change, wine becomes vinegar, seeds grow into trees, and the bust of

Socrates loses an ear. Some changes are accidental in the sense that they involve no substantial change. The Socrates who grew old, fat, and bald was not the same young, slim, hirsute youth who married Xanthippe. Something, however, had remained constant throughout these changes, and it was this which was seen as the substantial form. There was clearly something which made a dog a dog, a cat a cat, and Socrates, despite changes in his appearance, a man – namely the possession of canine, feline, and human substantial forms respectively. In addition, of course, individual dogs, cats, and men would possess numerous distinctive accidental features or forms.

The temptation to use the concept of substantial form to explain natural phenomena was irresistible. Most famously of course it appeared in the *virtus dormitiva*, or dormative power, Molière attributed to opium in his *La Malade Imaginaire* (1673). The reality behind Molière's parody can be seen in the complaint of Robert Boyle against his contemporaries raised in his *Origin and Form of Qualities* (1666):

For if it be demanded why jet attracts straw, rhubarb purges choler, snow dazzles the eyes rather than grass, etc. to say, that these and the like effects are performed by the substantial forms of the respective bodies, is at best to tell what is the agent, not how the effect is wrought.[19]

Presumably Boyle had in mind scholars like Puteanus, who was quoted by Gilbert as explaining the operations of the lodestone in terms of 'its substantial form . . . its own most potent nature and its natural temperament'.[20]

The issue was taken up by other philosophers of the period. Locke confessed in the *Essay* that he could make no sense of the notion. When told that bodies contained 'besides figure, size, and posture of the solid parts . . . something called *substantial form*', he could only respond that he had no idea of any such 'form'.[21] Descartes also found the notion obscure. It was quite 'unintelligible', he argued, to suppose that 'these qualities or forms could have the power subsequently to produce local motions in other bodies.'[22] The most damaging treatment, however, remained that of Boyle. In an unusually terse argument he dismissed such forms on the grounds that 'whatever we cannot explicate without them, we cannot neither intelligibly explicate *by* them.'[23]

Four elements

Numerous theories of elements had survived the medieval period and were widely deployed among renaissance scholars. The most common view, endowed with Aristotelian authority, took the four elements – earth, air, fire, and water – as the basic constituents of the physical universe.[24] They were not, of course, meant to be taken too literally. This would have made life far too hard. Instead, anything remotely liquid could be passed off as a watery element, and anything solid could be taken as earth. With such loose categories available, the working chemist would be most unlikely ever to face any genuine difficulties. Anything presented to him could readily be shown to be composed of some combination of the four elements.

Clearly, once again, difficulties were obscured, and problems buried within the over-generous categories provided by element theories. Why is mercury liquid? Because, the Aristotelians answered, 'it participates much of the nature of water.' And why is it so heavy? ''Tis by reason of the earth that abounds in it.' The analysis was submitted to a devastating critique by Boyle in his *The Sceptical Chymist* (1661). The account of mercury, he argued, made no sense. If the heaviness of mercury was due to the presence of the heavy element, earth, how could mercury be heavier than the same bulk of earth?[25]

Nor, he went on to argue, did such an approach explain anything. 'It was one thing to know a man's lodging, and another, to be acquainted with him.' That is, to be told that something is heavy because it contains earth is to be told no more than that a quality resides in an element. Also required is 'the cause of that quality, and the manner of its production and operation'. The Aristotelians, he concluded, were 'incompetent to explicate the origin of qualities', and would remain so 'while they endeavour to deduce them from the presence and proportion of such and such material ingredients'.[26]

Boyle's objections derive more from methodology, from assumptions about the nature of explanation, than from chemistry. He was equally clear about what constituted a sound explanation and, consequently, with what he wished to replace the Aristotelian account of nature. 'For tis by motion that one part of matter acts upon another', he declared, as he prepared to substitute his mechanical philosophy for the besieged peripatetic philosophy.[27] He chose to state his case in terms of that favourite example of the mechanical philosophers, the clock. We do not gain all that much understanding by being told that the weights of a clock are made of lead, the wheels of brass, and the hands of iron. The clue to the clock's operation lies in its mechanism, the manner in which weights drive wheels, and wheels drive hands.[28]

The four causes

Renaissance scientists inherited from Aristotle the notion that there are four kinds of cause – efficient, material, formal, and final. The major difficulty with this analysis concerned the issue of final causes. Once allow events and processes to be analysed indiscriminately in terms of their ends, their final purpose, and yet again over-generous patterns of explanation had been introduced into science. Some processes, of course, especially biological ones, were best understood in terms of their purpose. What role do gills play in fish? Or hair in mammals? How, other than by considering the ends served by the organs, could such questions be answered?

Yet, extend the use of final causes to more physical processes, to motion for example, and immediately the issue becomes less clear. An apple falls to the ground. What end is served here? Or, what is the final cause involved in the movement of arrows through the air, planets around the sun, or in the movement of iron filings to the lodestone?

Two kinds of answer were possible. The first was merely verbal and simply assumed that because an object regularly behaved in a certain manner it was thereby fulfilling its end. The argument that iron is attracted to lodestone because it is the end of iron to behave in this way is clearly no more satisfactory than the attribution of a dormative power to opium.

The alternative approach, adopted by Aristotle, involved the construction of a general theory of motion in which there was a genuine role for final causes to play. Thus, Aristotle distinguished between natural and violent motion. Terrestrial bodies moved naturally in a straight line, either upwards, as with air and fire, or downwards in the manner of water and earth. The reason apples fall directly to the ground is because they thereby move as naturally as possible in a straight line to earth. Unconstrained and left to themselves all bodies moved directly and rectilinearly to their natural places. Why then do apples fall? Because it is the end of earthly bodies to move to their natural place, namely, earth. Bodies, like arrows shot from a bow, which seemed to violate this account, were held to move violently. In this and other cases of violent motion bodies are pushed or pulled away from their natural paths. The natural motion of heavenly bodies, however, was circular, considered to be more perfect than rectilinear motion.

Thus, final causes could find a place in a rationally worked out physics of motion. The disadvantage, of course, was that they could not be understood outside the system. And, indeed, there were few willing to accept the system in its entirety. What, for example, were the pushes and pulls which kept projectiles flying through the air? And if the crucial distinction between natural and violent motion broke down, so also did all talk of the ends of motion.

The price for final causes could thus often prove to be too high. Descartes would have none of them and sought to expel them from science completely. Accordingly, he noted in his *Principles of Philosophy* that 'It is not the final but the efficient causes of created things that we must enquire into.' To do otherwise would be to assume that 'we can share in God's plans.'[29] Boyle, also, shunned all arguments which suppose 'in nature and bodies inanimate, designs and passions proper to living and perhaps peculiar to intelligent beings'.[30]

It has become apparent that while the philosopher-scientists of the seventeenth century were busily disposing of the last remaining residues of Aristotelian thought, they were at the same time constructing the outlines of the more powerful mechanical philosophy. But before this could become widely accepted a further challenge to science, the reappearance of scepticism in Europe, had first to be surmounted.

Scepticism

In 1562 Henri Estienne published the first modern translation of Sextus Empiricus' *Outlines of Pyrrhonism*. Almost unknown in the medieval period, the *Outlines* dates from the second century AD and is an extensive account of the sceptical tradition as it had developed in antiquity. For many at that time, and since, Estienne's publication was the beginning of modern philosophy. Pierre Bayle in his *Dictionnaire* (1696) so described it, while Annas and Barnes more recently have judged that it was 'the rediscovery of Sextus Empiricus and of Greek scepticism which shaped the course of modern philosophy for the next 300 years'.[31]

Sextus actually presented ten modes or tropes. In each case variations in the subject doing the judging, or the object being judged, led to uncertainties. The general form of the argument, as presented by Annas and Barnes, is as follows:

x appears F to H.
x' appears F' to H'.
There are no grounds to prefer either H or H'.
Therefore, we must suspend judgement whether x is an F or an F'.

For example, Demophon, Alexander's waiter, unlike others, shivered when he was in the sun and felt warm in the shade. Consequently, we must suspend judgement on whether the day was warm or cold. And so on, through all the tropes, we must finally conclude that we must suspend judgement on all issues.

The revival of Greek scepticism in the sixteenth century has been charted by Richard Popkin in his well-known work *The History of Scepticism from Erasmus to Spinoza*. He notes that one of the earliest of the modern sceptics, Francisco Sanchez (1552–1623) in his *Quod nihil scitur* (1581), argued specifically that scientific knowledge was impossible. All routes were surveyed and found to be deficient. Demonstration was rejected because it is syllogistic, and syllogisms were shown to be circular. How can we conclude that 'Socrates is mortal' from the premises, 'All men are mortal', and 'Socrates is a man', without first having shown that Socrates is indeed one of the men who is mortal? Nor can anything be gained by studying causes, as this leads to the inevitable regress of searching, never-endingly, for causes of causes of causes. The only knowledge Sanchez would allow is of particular objects present in our experience.[32]

More influential, and certainly better known, were the works of

Montaigne. As he wrote his sceptical tract *Apologie de Remond Sebond*, Popkin notes, he had slogans and phrases from Sextus Empiricus 'carved into the rafter beams of his study, so that he could brood upon them'. Science, he noted, had little to offer. Ptolemy had proclaimed the universe to be earth-centred, Copernicus that its centre was the sun, and no doubt, Montaigne added, a new theory will appear one day to overthrow both of these.

Scientists of a more recent period, with greater confidence in their powers, can afford to ignore the sceptic's threat. Do not their telephone systems and missions to Uranus prove the absurdity of the Pyrrhonist (completely sceptical) position? Scientists of the early seventeenth century had no such triumphs with which to justify their position. Nor was it then apparent that science would ever attain successes of this kind.

The initial philosophical response to scepticism, linked with Gassendi and Mersenne, is described by Popkin as constructive scepticism, a compromise in fact between scepticism and dogmatism. The truth of the Pyrrhonist position was conceded, and it was accepted that there could be no real knowledge of nature. At the same time, it was claimed that even if we could have no sure knowledge of reality and ultimate causes, we could at least have some knowledge of effects and appearances. Such was the argument of Mersenne's *La vérité des sciences contre les sceptiques* (1625). In all fields, he insisted, we find genuine knowledge. There are truths like 'the whole is greater than its parts', and 'an elephant is bigger than an ant', known to us all and indisputable. Mersenne, in fact, Popkin reports, devoted 800 pages of his work to listing those matters on which we need not suspend judgement.[33]

Mersenne's position is clearly unsatisfactory, granting either too much or too little to the sceptic. Once accept part of the sceptic's position, and then, as Descartes clearly saw, it is impossible to deny his claims in full. To emphasise the point Descartes set about strengthening the sceptic's argument to such a degree that it could easily absorb Mersenne's counter-examples. There was little point in considering any modified form of scepticism; Descartes wanted to refute it at its strongest point. Indeed, he sought to find, at this point of greatest strength, its ultimate weakness.

In his first statement of the problem in his *Discourse on Method*

(1637), Descartes proposed in Part IV to reject 'as if absolutely false everything in which I could imagine the least doubt'. Since our senses deceive us, and men make mistakes in reasoning, and are deceived by dreams, he resolved to pretend that everything he had ever thought was false. Yet even this programme seemed too mild to the Descartes of the later *Meditations* (1641). In the first *Meditation* he added to the sceptic's case a malignant demon, exceedingly potent and deceitful, who employed all his skill to confuse Descartes. Whereas Descartes in 1628, in his *Regulae*, had allowed arithmetic and geometry to be 'free from any taint of falsity or uncertainty', the *Meditations* permitted no such concession.

But, having taken scepticism as far as it could go, having granted it all possible concessions, Descartes set about showing in the second *Meditation* how science could once more be established. For, he argued, even though he could doubt the existence of his own body, he could not doubt his own existence. Indeed, the very fact that he was deceived, granted his existence: 'Doubtless, then, I exist, since I am deceived.' And, of course, the harder questions of doubt are pressed, the firmer becomes the proof of Descartes's own existence. At this point, having established the one indubitable point in his system, Descartes continued throughout the *Meditations* to reconstitute modern science.

The case of China

Everything considered so far has been taken from western scientific and philosophical traditions. There is, however, thought outside the West. It will be instructive to see how science and philosophy have developed in at least one non-western tradition.

The issue is one of the major themes pursued by Joseph Needham in his still continuing *Science and Civilisation in China*. In volume 2 in particular, *History of Scientific Thought*, he attempted to trace some of the interactions between Chinese science and the numerous schools of Chinese philosophy. For the Confucians, he noted, as with Socrates, the only proper study of mankind was man. Although they were rationalists, and opposed to all forms of superstition, they were concerned exclusively with human social life. Non-human phenomena were consequently ignored. Even the dead were discounted.

Asked how man can serve the spirits, Confucius characteristically replied: 'You do not yet know about the living, how can you know about the dead?'[34] Only four subjects were considered suitable for study: culture, the conduct of affairs, loyalty to superiors, and the keeping of promises. Unworthy of study were such disorders of nature as comets and volcanoes.

Nor were the Confucians any more tolerant of theoretical thought. Needham quotes from Hsun Tzu, a third-century Confucian, a passage listing all the things which have nothing to do with the distinction between right and wrong. They are, he commented, 'things the knowledge of which does not benefit men, and ignorance concerning which does no harm . . . They belong to the speculations of unruly persons of a degenerate age.'[35] Being ignorant about 'the displacement of body and empty space', Hsun Tzu concluded, does not make us lesser gentlemen. Without such knowledge 'artisans can be just as good artisans', and rulers just as good rulers.[36] With such a philosophy, Needham noted, 'there was no room for science . . . only traditional technology.'

What of the rival philosophical tradition of Taoism? Its roots were two. Support came first from the philosophers who had fled the courts of the feudal princes, concerned obsessively with rank and ritual, for the wilderness where they could meditate upon the order of nature. The second root of Taoism lay within the tradition of the ancient shamans and magicians. The way or Tao both groups sought was concerned much more with the manner in which the universe worked than with the proper ordering of society. At first sight it would therefore seem that such views would be much more favourable to the development of science.

Indeed Needham does find in Taoism many factors favourable to the development of science. It had, like science, universal pretensions. The Tao, Lao Tzu insisted, is 'everywhere', in ants, weeds, and even dung.[37] Just as nothing is outside the domain of science, Needham argued, nothing was too repulsive or trivial for the Tao. Further, Taoism was concerned with the discovery of the causes of things. Thus, from a third century BC Taoist sage, Needham quotes: 'All phenomena have their causes. If one does not know these causes . . . it is if one knew nothing, and in the end one will be bewildered.'[38] And, thirdly, the Taoists adopted an empirical approach to nature. Unlike the Confucians they were reluctant to impose their will on anything. They sought, rather, to allow things 'to work out their destinies in accordance with their intrinsic principles'.[39] But to fulfil this approach, it was first essential to know how things developed naturally. Hence the concern with the careful study of nature.

The philosophical tradition of Tao did, Needham has repeatedly insisted, lead to the development of science in China. And in Ko Hung, for example, a Taoist of the fourth century AD, he claims to identify 'scientific thinking at what appears to be a high level'.[40] Yet, despite such beneficial influences, and despite a number of early successes, science in China failed to develop. There was, in short, no scientific revolution in China.

The reasons for this are, no doubt, numerous and complex. One intriguing comment on the situation has come from the sinologist and historian of science Nathan Sivin. Although China undoubtedly produced any number of astronomers, chemists, mathematicians, and physicians, he has argued, it produced no *scientists*. The concept of science simply did not emerge in China. According to Sivin:

There does not seem to have been a systematic connection between the sciences in the minds of the people who did them. The sciences were not integrated under the domain of philosophy, as schools and universities integrated them in Europe and Islam. Chinese has sciences but no science, no single conception or word for the overarching sum of all of them.[41]

The words normally taken to denote science, *hsueh*, and *li*, refer respectively to, Sivin claims, everything we can learn through study, and any comprehensible pattern.

This failure may well, in turn, have arisen from the earlier failure of Chinese philosophy to develop an appropriate concept of knowledge. The Confucians dismissed outright any kind of theoretical knowledge, while the Taoists reduced much of its power by transforming it into a form of esoteric knowledge. Lacking the western notion of knowledge, Chinese thought was also forced to do without the corresponding western notion of science.

ARE SCIENTIFIC THEORIES RATIONAL?
ARE THEY TRUE?

Since Ptolemy was once mistaken over his basic tenets, would it not be foolish to trust what moderns are saying now?
Montaigne, *An Apology for Raymond Sebond* (1580)

A Parisian is taken aback when he is told that the Hottentots cut off one testicle from their male children. The Hottentots are perhaps surprised that the Parisians keep two.
Voltaire, *Philosophical Dictionary* (1764)

Now how does one alter the charge on the niobium ball? 'Well at that stage', said my friend, 'we spray it with positrons to increase the charge or with electrons to decrease the charge.' From that day forth I've been a scientific realist. So far as I'm concerned, if you can spray them then they are real.
Ian Hacking, *Representing and Intervening* (1983)

From the seventeenth century onwards it has become increasingly clear to many that science offers the only rational approach to the study of nature. Such a view could be seen, for example, in the increasingly confident manner in which writers began to dismiss the claims of tradition in favour of the newer methods of science. Pierre Bayle, author of the influential *Dictionnaire* (1696), made the point that 'It is the purest delusion to suppose that because an idea has been handed down from time immemorial . . . it may not be entirely false.'[1] He also spoke of the 'oracle of reason' before which the superstitions of the past would collapse. A similar point was made by Halley in the

'Ode' he contributed to Newton's *Principia* (1687). Contrasting the learning of the ancients with the thought of his own day, Halley concluded:

> Matters that vexed the minds of ancient seers,
> And for our learned doctors often led
> To loud and vain contention, now are seen
> In reason's light, the clouds of ignorance
> Dispelled at last by science.

By the nineteenth century the attractions of science had begun to appear boundless. History itself came to be presented as the inevitable advance towards science. Auguste Comte, the founder of positivism, proposed a scheme of historical development in his *Philosophie positive* (1830–42). Three ages were recognised. In the first age, the supernatural, all events were explained as the effects of supernatural powers with, for example, thunder being seen as the wrath of Zeus. In the second, metaphysical age, abstract forces replaced gods, and the same thunder would be explained by appeal to nature's fear of the void. In the final and highest stage, scientific positivism, the search for causes has been abandoned. In their place scientists simply seek correlation between events. Consequently, rather than concern themselves with the cause of thunder, scientists turn instead to discovering those meteorological conditions invariably linked with thunder.

Tripartite divisions of history, ever popular, followed in profusion.[2] The anthroplogist Sir James Frazer, for example, in his widely read *The Golden Bough* (1890), argued that history has passed through the three ages of magic, religion, and science. It was also becoming apparent that rates of development were far from uniform. There was much talk of the peoples in newly colonised parts of the world as primitive man, still in the age of magic, supposedly lacking the ability to generalise and to think logically. Even their languages were suspect and it was seriously claimed that some people were incapable of communicating at night, when their hand signals and facial expressions could no longer be seen.[3]

Views of this kind have been recognised by later scholars as simple nonsense. Yet, while notions of primitive mentality were gradually abandoned, science and reason continued to be identified. Nowhere has this identification been more vigorously espoused than in the writings of Popper. In his *Back to the Presocratics* he has insisted that 'the critical and rationalist tradition was invented only once', by the Presocratic philosophers, and 'rediscovered and consciously revived in the Renaissance, especially by Galileo'. The tradition represents 'the only practicable way of expanding our knowledge'.[4] The tradition, of course, once established is open to all, and it is assumed that when it is revealed to the members of a 'primitive' society, magic, superstition, and other irrational procedures will be immediately dropped in favour of the methods of science. How could spells and incantations compete with fertilisers and tractors? And how could fetish priests and traditional healers survive alongside physicians and engineers?

Evans-Pritchard and the Azande

The first major challenge to these assumptions came from a most unlikely source – Evans-Pritchard's seminal *Witchcraft, Oracles and Magic among the Azande* (1937). From 1926 to 1930 Evans-Pritchard, then a young anthropology lecturer at the London School of Economics, worked among the Azande in the southern Sudan. Detailed studies were made of their history, politics, economy, and, most significantly of all, their thought. They revealed a society dominated at all levels, technological as well as theological, by a belief in witchcraft. The central feature of the system was that all unexpected events – accidents, deaths, illnesses, crop failures, storms, or whatever – were attributed to the behaviour of witches. A complex system of oracles enabled the witches to be identified, and vengeance and compensation pursued.

In the course of these studies Evans-Pritchard established the following general points about the Azande belief system.

1 Witchcraft belief was ubiquitous:

> It plays its part in every activity of Zande life; in agriculture, fishing and hunting pursuits; in domestic life . . . in communal life . . . and court . . . there is no niche or corner of Zande culture into which it does not twist itself. If blight seizes the groundnut crop it is witchcraft, if the bush is vainly scoured for game it is witchcraft . . . if a wife is sulky and unresponsive . . . it is witchcraft . . . if . . . any failure or misfortune falls upon anyone at anytime . . . it may be due to witchcraft.[5]

2   It was a normal and accepted part of life. Westerners do not become 'psychologically transformed' on hearing that someone is ill or that a car failed to start; nor do the Azande respond differently when they hear of a case of witchcraft. 'They expect people to be bewitched', he noted, 'and it is not a matter for surprise or wonderment'.[6]

3   Seen from outside and as a whole the system is not without contradiction. For example, oracles sometimes give contradictory answers to the same question. Asked whether a particular individual is a witch, they first answer 'Yes', and subsequently 'No'. Or again, it is claimed that witchcraft is inherited by unilineal descent from parent to child. Thus the sons of a male witch should all be witches. Yet when a man is demonstrated to be a witch his male descendants invariably deny that they are also witches.[7]

4   Most significantly of all, Evans-Pritchard pointed out that the system, contradictions and absurdities included, was viable. Although he had found it strange at first to live in such a community, he discovered that 'After a while I learnt the idiom of their thought and applied notions of witchcraft as spontaneously as themselves in situations where the concept was relevant.' And, like any Azande, Evans-Pritchard soon found himself consulting oracles and regulating his life in accordance with their advice. It was, he concluded, 'as satisfactory a way of running my home and affairs as any other I know of'.[8]

The contrast between the work of Evans-Pritchard and his predecessors was considerable. 'We are red macaws', the Bororo of Central Brazil told Von Steinem in 1894. The claim was taken seriously by scholars of the period. 'The Bororo sincerely imagines himself to be a parrot', Durkheim and Mauss informed their readers in 1903.[9] Clearly, people who thought themselves to be parrots occupied a completely different intellectual world to those who saw parrots merely as members of the order *Psittaciformes*. Yet Evans-Pritchard presented Azande thought about the natural world as a coherent and workable belief system. Presumably detailed examination of Bororo culture would reveal an equally coherent system.[10] Science, for the first time, no longer seemed to be the only rational account of nature; it was surrounded if not by rivals, at least by companions. No longer could the beliefs of the Azande, nor for that matter of the Ashanti, Yoruba, Tikopia, Hopi, and many others, be dismissed as illogical or absurd; they were in fact as well formed and as sensible as any western way of life.

Horton's advance

Evans-Pritchard had managed to show no more than that the Azande belief system operated satisfactorily at a practical level. Similar studies by Malinowski on the canoe and garden magic of the Trobriand Islanders, and by Firth on the Tikopia economy, revealed similar states of affairs.[11] Despite a reliance on magic, Trobriand gardens flourished and their canoes sailed long distances. In such activities they were doing something readily comparable with western technicians. They may not have been as efficient as their western counterparts, but the spells they used to gain a good harvest were part of the same practical culture as the fertilisers used by the farmers of Europe. Scientists, however, created theories as well as practical effects. And while many could recognise the practical achievements of traditional societies, they still reserved for western science a monopoly of serious theoretical thought about nature. The rest was superstition, and while it was conceded that holding such irrational beliefs could well serve a number of vital social functions, the beliefs themselves could never form part of a coherent intellectual system.

The first full statement of an alternative position came in Robin Horton's much discussed 'African Traditional Thought and Western Science', first published in 1967.[12] Anthropologists, he claimed, had long failed to understand the true nature of traditional religious thought largely because they were unfamiliar with the theoretical thought of their own culture. Theoretical thought in science, Horton claimed, involves 'the elaboration of a scheme of entities or forces operating "behind" or "within" the world of common-sense observations'.[13] In this manner scientists explain such observable phenomena as 'a large mushroom cloud on the horizon' by reference to the fusion of numerous unobservable and theoretical hydrogen nuclei. And so on, through most parts of science, where explanations are likely to be made by referring to the operation of such theoretical entities as electrons, genes, viruses, gravity, and the unconscious.

The traditional thinker, Horton insists, argues in a similar manner, only the idiom varies. The events of his life and of nature are likely to be governed by the operation of various spirits and gods. These are as explanatory and as theoretical, just as much a link between theory and experience, as are the quanta of modern physics. Consequently, Horton emphasised:

> To say of the traditional African thinker that he is interested in supernatural rather than natural causes makes little more sense, therefore, than to say of the physicist that he is interested in nuclear rather than natural causes. Both are making the same use of theory to transcend the limited vision of natural causes provided by common sense.[14]

First response

The natural response is to insist that Azande and other belief systems, though coherent, are simply false. The claim that syphilis is caused by witches is as baseless as the belief that it can be acquired from toilet seats. The scientific response, however, that the disease is spread by the spirochaete *T. pallidum*, is a true account. Once the criterion of truth is ignored, then indeed coherent conceptual schemes can be multiplied almost endlessly. Nor is it necessary to travel to the Sudan or the Amazon to encounter them; they can be found just as readily in the pages of Homer, St Augustine, or Jonathan Swift. Once allow the concept of truth, along with the related notions of proof, verification, and refutation, and science will once more occupy a central and unique position.

The argument fails, however, once it is realised that science has a history. And it is a history rich with discarded and false theories. Ptolemaic astronomy, Cartesian mechanics, and Linnean systematics are all, along with countless other theories, as *coherent* and *false* as Azande witchcraft beliefs. The people of Baan Phraan of Thailand deny that crows are birds, and the Karam of New Guinea take a similar view of the cassowary.[15] These views are undoubtedly false. Are they, however, any more false than the claim made by Linnaeus in 1758 that bats were primates, or his earlier judgement of 1735 that whales were fish, spiders were insects, and pelican were geese?[16]

Nor are errors of this kind absent from contemporary science. The first edition of the Penguin *A Dictionary of Biology* (1951) under the entry 'Chromosome' notes that 'man has twenty-four pairs'. Human chromosome-counting had begun in the 1890s and, as with any new technique, early results were variable and inaccurate. One reason for this was the practice of cytologists of basing their counts on cells taken from cadavers. Such cells, apparently, clump together rapidly and present misleadingly low counts. In 1923, however, T. S. Painter, working with the testes of three recently castrated patients, established a consistent figure of forty-eight chromosomes. For thirty years Painter's work was repeatedly confirmed by a whole generation of cytogeneticists.

New techniques became available in the early 1950s which allowed the cell nucleus to be examined with much greater clarity. Whereas Painter could see something looking like a plate of spaghetti, these new techniques allowed the chromosomes to be seen clearly separated, and spread out in the same optical plane. Soon after, in 1956, Tijo and Levan announced that human cells contained only forty-six chromosomes. Consequently the fourth edition of the Penguin *A Dictionary of Biology* (1961) noted that while *Drosophola melanogaster* continued to possess four pairs of chromosomes, man had lost a pair since 1951 and now possessed only twenty-three. The figure is repeated in the seventh edition of the *Dictionary* (1980). Whether or not further revisions will be required in the future remains to be seen.[17]

Attempts to define science in terms of truth are likely to prove to be far too costly. They may well succeed in distinguishing between science and traditional thought, but only at a price of radically separating science from its own past.

The resort to method

Eighteenth-century disciples of Newton had to face the possibility that, in the words of William Emerson, 'The Newtonian philosophy, like all others before it, will grow old and out of date, and be succeeded by some new system.' The objection, Emerson responded, 'is falsely made':

For never a philosopher before Newton ever took the method that he did. For whilst their systems are nothing but hypotheses, conceits, fictions, conjectures, and romances, invented at pleasure and without any foundation in the nature of things, [Newton], on the contrary . . . admits nothing but what he gains from experiments and accurate observations . . . It is therefore a mere joke to talk of a new philosophy . . . Newtonian philosophy may indeed be improved and further advanced; but it can never be overthrown.[18]

Emerson's devotion to Newton may have been extreme, and his optimism misplaced; his general argument, however, has continued to attract support.

The theories of science, however abstract and speculative they may be, whether true or false, always, at some point, confront experience directly. The essence of scientific method in the eyes of Emerson and others is to admit nothing which does not ultimately derive from 'experiments and accurate observations'. The method, of course, is not quite as simple as this and further rules are needed to allow scientists to determine whether or not their theories are supported or refuted by some particular experiment or observation. The rules themselves may be complex, and not always easy to apply. Nevertheless, the central point remains: namely that the theories of science are tested against experience and only continue to be accepted in so far as they survive these tests.

It might at first sight be thought that all belief systems have to face up to nature. Only sheer fantasy can be allowed to ignore completely the constraints of reality. Evans-Pritchard and other anthropologists, however, have shown this assumption to be naïve. Azande thought, Evans-Pritchard argued, was characterised by an extensive use of what he termed to be secondary elaborations. By this he meant that whenever Azande beliefs and reality were about to conflict, a secondary assumption was introduced to restore the original harmony.

For example, let us return to the apparent contradiction introduced earlier. Witchcraft is inherited unilinearly, from father to son, and mother to daughter. How, therefore, can I accept that my brother is a witch and yet deny that I am also infected? To prevent this absurdity arising, the Azande adopt further 'elaborations of belief'. They argue, for example, that 'If a man is proven a witch beyond all doubt his kin, to establish their innocence . . . deny that he is a member of their clan. They say he was a bastard, for among Azande a man is always of the clan of his genitor and not of his pater.'[19] In this and other ways, Evans-Pritchard concluded, the Azande freed themselves from 'the logical consequences of belief in biological transmission of witchcraft'.

Compare this with one of the great demonstrations of modern science. Einstein's General Theory of Relativity is for many as implausible and as unfamiliar as Azande witchcraft notions. Yet Einstein's theory differed in one crucial respect: it brought with it a number of precise, specific, and unexpected predictions. One such was the claim that light would be deflected by the presence of a strong gravitational field. Starlight grazing the sun should therefore be displaced by a measurable amount. This was calculated by Einstein to be 1.17″ of arc. On 29 May 1919 a total eclipse of the sun would be observable from Principe, off the West African coast. It was observed by Sir Arthur Eddington. When Eddington came to analyse his material, described by him as the greatest moment of his life, he found the results to agree with Einstein's predictions.

The impact of Eddington's demonstration was long remembered by a number of impressionable young scientists and philosophers. They were struck by the unparalleled manner in which science could be both so bold in its claims, yet so careful and strict in testing those claims. Thus Popper, writing in 1957, recalled how in 1919 'We all . . . were thrilled with the result of Eddington's eclipse observation . . . It was a great experience for us, and one which had a lasting influence on my intellectual development.'[20] Another who, many years later, recalled the drama of the event was the astronomer Cecilia Payne-Gaposchkin, then a young Cambridge student. She actually heard Eddington announce his results in the great hall of Trinity College:

The result was a complete transformation of my world picture. I knew again the thunderclap that had come from the realization that all motion is relative . . . For three nights, I think, I did not sleep. My world had been so shaken that I experienced something very like a nervous breakdown.[21]

From this, it has been argued, the sovereignty of science can be seen to require no secondary elaborations; its strength and durability come from the manner in which, even at its most abstract and theoretical, science can still be publicly confirmed by experience.

The Kuhnian reply

Much of the philosophy of science of the last twenty-five years has been directed against this view. The first major attack was developed by Thomas Kuhn in his *The Structure of Scientific Revolutions* (1962). Kuhn's main argument is now very familiar. In brief, he drew a distinction between normal and revolutionary science. Normal science is essentially puzzle-solving and takes place within a well- and often long-established paradigm. At certain periods the paradigm will break down. Puzzles will resist solutions, anomalies will multiply, and the most creative workers, instead of exploring the depths of the prevailing paradigm, will seek to overthrow it. At this point a new paradigm is likely to emerge and consequently allow scientists to return to their traditional practice of puzzle-solving within the parameters of a well-established paradigm. In this manner, after trying for centuries to fit stellar observations into the framework of the Ptolemaic paradigm, and after an appropriate period of revolutionary activity, astronomers began in the seventeenth century to fit the same data into the heliocentric paradigm of Copernicus.

In addition Kuhn argued for a number of subsidiary theses. Among these the following are particularly relevant:

1 'More than one theoretical construction can always be placed upon a given collection of data.'
2 Scientists, 'when confronted by even severe and prolonged anomalies . . . do not renounce the paradigm that has led them into crisis'.
3 'No process yet disclosed by the historical study of scientific development at all resembles the methodological stereotype of falsification by direct comparison with nature.'
4 When presented with anomalies scientists 'devise numerous articulations and ad hoc modifications of their theory in order to eliminate any apparent conflict.'
5 Propositions of the same kind as Newton's second law, and the chemical law of fixed proportions, behave so much 'like a purely logical statement that no amount of observation could refute'.
6 'Failure to achieve a solution discredits only the scientist and not the theory.'[22]

Substitute witchcraft beliefs for science, and the Azande for scientists, and the above six propositions could be taken from the pages of Evans-Pritchard. Indeed, the supposed similarity had been noted earlier by the physical chemist and philosopher Michael Polanyi. Both Azande and scientific beliefs were, he recognised, stable. The Azande maintained the stability of their beliefs by suppressing alternative views. Science, he argued, did the same. He described from his own scientific career the fate of the theory of electrolytic dissociation introduced into chemistry in the 1880s. Although it gave good results with weak electrolytes, the theory could do nothing for such strong electrolytes as salt and sulphuric acid. The discrepancy between theory and experience was noted in textbooks for thirty years. The existence, however, of anomalous strong electrolytes was never taken to refute the theory. When in 1919 Polanyi first came across the proposal it was received by him with 'amazement'.[23]

Support for Kuhn's claims came from a number of sources. The most impressive evidence came from a large number of detailed historical studies of major scientific advances. The works of Galileo, Dalton, and Mendel were examined amongst others; in all cases theory seems to have been preferred to experimental data. Galileo in his *Two New Sciences* (1638) described how he came to discover the law of free fall, that the spaces traversed by a falling body were to each other as the squares of the times. A piece of wooden moulding 12 cubits long with an incised groove along its entire length was placed in a sloping position; balls were rolled down it from various heights and their descent timed. The system for timing consisted of a large water vessel from which a water jet escaped, the water being collected below in a small glass during the ball's descent. The water was then weighed in a balance, and the ratio of their weights gave the ratio of the different times of descent of each ball. In this manner Galileo was able to establish such propositions as that the ratio between the time a ball rolls a quarter-length and the whole length was 'precisely one-half'. The ratio, Galileo noted proudly, was 'precisely that ratio . . . the author has predicted'.[24]

The historian of science, Koyré, could scarcely conceal his incredulity:

A bronze ball rolling in a 'smooth and polished' wooden groove! A vessel of water with a small hole through which it runs out and which one collects . . .

in order to weigh it afterwards . . . what an accumulation of sources of error and inexactitude! It is obvious that the Galilean experiments are completely worthless: the very perfection of their results is a rigorous proof of their incorrection.[25]

What of John Dalton, the founder of modern atomism? Dalton had supposedly tested his atomic hypothesis by analysing the various nitric oxides, and found in them confirmation of his views. A modern chemist, however, has reported a different result: 'From my own experiments I am convinced that it is almost impossible to get these simple ratios in mixing nitric oxide and air over water.'[26]

Mendel's work on sweet peas has received a similar modern reappraisal by the statistician R. A. Fisher: 'The data of most, if not all, of the experiments have been falsified so as to agree closely with Mendel's expectations.'[27] It is today widely accepted that Mendel first worked out his theory and consequently adjusted his experimental data to fit his initial theoretical expectations.

Numerous other examples could be produced. Newtonian physics, Ptolemaic astronomy, and nuclear physics have all been surveyed in recent years and have shown that even the most eminent of scientists have been prepared to modify observational and experimental data to satisfy the needs of theory.[28]

We have already seen the apparent indifference of scientific theory to unfavourable evidence. If the phenomena predicted by theory fail to appear, their absence may simply show that the search for them has been insufficiently thorough. For this reason high energy physicists continue their search for such elusive entities as the magnetic monopole, gravitational waves, the Higgs boson, and an unstable proton.[29] Consider a more remote example. As is well known, the geologist Charles Lyell argued strongly in his *Principles of Geology* (1830–3) for a radical uniformitarianism. This meant that he refused to explain terrestrial geology in terms of past catastrophes. Such views were shared by many of his contemporaries. Lyell, however, went further and claimed that his uniformitarianism was incompatible with the progressivism equally favoured by his colleagues. The earth, Lyell insisted, had always appeared the same. But what of the fossil record? Did not this clearly show successive ages of fish, reptiles, birds, and mammals? Why, for example, did the Devonian period reveal ample fish remains, but not a single mammalian fossil? Lyell replied that fossils are normally found in marine strata because they are best preserved there. A mammal, however, washed into the sea would probably be eaten. The few escaping this fate, preserved in the sea-bed of some ancient ocean, are likely to remain outside the range of normal geological expeditions for a long time. So Lyell found a 'good' reason to ignore the evidence and keep to his own theory.

In a similar way, the appearance of unwanted entities need not prove to be unduly embarrassing to the modern scientist. Impure samples, contaminated equipment, procedural errors, and many other causes routinely produce strange results in even the best regulated laboratory. Consequently unpredicted phenomena are more likely to be dismissed as some trivial aberration than as a challenge to established theory. For example, in 1896 Henri Becquerel discovered the new and unexpected property of radioactivity. The discovery is invariably described as accidental. Becquerel just happened to have in the same drawer a sample of uranium salts and a box of photographic plates. To his surprise he found that uranium within a dark drawer could produce an image on plates wrapped in two sheets of thick black paper. Further work convinced Becquerel that he had indeed discovered a genuine effect. Yet before Becquerel there must have been many other scientists who found fogged photographic plates mysteriously appearing in their laboratories. One worker, tradition records, weary of his assistant's complaints that photographic plates persistently fogged over when kept with uranium, even gave instructions that the plates and uranium samples be kept in separate drawers, but did not investigate the phenomenon.

Much of this seems remarkably similar to many of the points made by Evans-Pritchard. Secondary elaborations of belief, it seems, can be identified as readily in the laboratories of Europe as in the villages of the Azande. Compare, for example, the following statements of Kuhn on modern science with the earlier claims of Evans-Pritchard on Azande thought:

Kuhn (1970)	*Evans-Pritchard* (1937)
1 'Newton's second law of motion behaves . . . very much like a purely logical statement that no amount of observation could refute' (p. 76).	1 'Since [magic's] action transcends experience it cannot easily be contradicted by experience' (p. 475).

2 'Failure to come near the anticipated result is usually failure as a scientist' (pp. 35–6).

2 'A witch-doctor is a cheat because his medicine is poor' (p. 194).

3 '[Scientists] devise numerous articulations and ad hoc modifications of their theory in order to eliminate any apparent conflict' (p. 78).

3 'Elaborations of belief free Azande from having to admit . . . the logical consequences of belief' (p. 24).

Realism

The notion that scientific theories are transitory phenomena rather than truthful descriptions of nature is not something that has recently emerged from the imagination of a school of eccentric philosophers. Jonathan Swift, for example, sent Gulliver on his third voyage to Glubbdubdrib, the isle of magicians, where he was invited by the governor to conjure up the spirits of his choice. Aristotle was summoned, along with Descartes and Gassendi. After freely admitting his own errors, Aristotle noted how Gassendi's atomism and Cartesian vortices had been rejected. He concluded by predicting a similar fate for 'Attraction': 'Whereof the present learned are such zealous asserters. He said that new systems of nature were but new fashions . . . and even those who pretended to demonstrate them from mathematical principles would flourish but a short period of time.'[30]

In more recent times the mathematician Henri Poincaré has also been struck by the 'ephemeral nature of scientific theories'. After a brief period of success, we see them 'abandoned one after another' and, consequently, we are tempted to conclude that 'the theories in fashion today will in a short time succumb in their turn.'[31]

If, indeed, theories of the past have been abandoned so regularly, then it would be most unlikely that the theories of our own day have any special claim to permanence or truth. We might as well suppose that 501 Levis will still be in fashion fifty years hence as that the theories of today will survive unchallenged. Authors of current textbooks proudly announce in their introductions how extensively they have been forced to revise their latest edition, seemingly unaware of the implications of their boast. A closer examination of the issues, however, shows that the case has been somewhat over-stated. Science is no more a matter of fashion than it is a system of timeless truths.

The general appeal to total truth clearly fails. It is possible, however, to mount a more limited defence of the role of truth in science. Much of the science of the past is undoubtedly false. This view is of course compatible with the claim that much is also true. And much of this material can be found happily presented in almost identical form in textbooks of the past and those of today. For example, the nineteenth-century chemist Henry Roscoe confidently described the 'hydrocarbons of the series C_nH_{2n+2}', the 'paraffin group', and listed amongst its members methane and ethane as CH_4 and C_2H_6 respectively.[32] Nearly a century later *The Penguin Dictionary of Chemistry* (1983) also refers to the paraffins by the general formula C_nH_{2n+2}, and goes on to describe methane and ethane as CH_4 and C_2H_6 respectively. Indeed, much of the material from chemistry textbooks of the late nineteenth century, often in a different terminology, can be found repeated in today's textbooks. It is also of course true that much more can be found in contemporary works. The point remains that while some theories change, much of science remains unaffected.

Secondly, many of the permanent features of science are judged more by their convenience and adequacy than by their truth value. For example, the ability to talk about and refer to the numerous visible stars of the heavens is very much part of science. The system was developed in the early seventeenth century by the German astronomer Johann Bayer. Whereas earlier generations of astronomers would refer clumsily to stars by such circumlocutions as 'the bright star in the left hand of Medusa', Bayer began the system by which each star can be uniquely identified by a combination of letters and numbers. And stars are, of course, just one aspect of nature which falls within the scientist's power to identify, name, describe, and classify. The rocks of the earth, chemical molecules, types of fibre, blood, or crystal structures, are all dealt with somewhere within the confines of the appropriate reference work.

Within these reference works much of the more permanent data of science are stored. Take, for example, the *International Critical Tables of Numerical Data: Physics, Chemistry, and Technology*. Here can be found listed, for many a chemical compound, its absorption spectra, boiling point, melting point, density, electrical conductivity, ionisation by X-rays, magnetic susceptibility, and much more besides. From a more specialised field there is the *Barker Index of Crystals*, providing an index of crystal angles arranged in numerical order. And from biology the *Index Kewensis* can be mentioned, a work which lists alphabetically the million or more generic and specific names used to identify plants since the time of Linnaeus.[33]

Thirdly, the errors of the past can be seen as early approximations to the truths of today, or, at least, to the truths of the future. What is important about the propositions and theories of science is not really whether they are true, but whether they are somehow advancing towards the truth. This advance can come, of course, just as much by the exposure of error as by the discovery of truth. As long as better methods are being developed, and misleading and inaccurate techniques discarded, it matters little that textbooks of the past were wrong about the atomic weight of hydrogen or the distance of the Andromeda nebula. What matters is that we now know how these erroneous results were obtained, and that we can advance systematically beyond them.

But what of scientific revolutions? Do they not introduce fundamental discontinuities into science? Against this view Whewell has argued:

The principles which constituted the triumph of the preceding stages of the science, may appear to be subverted and ejected by the later discoveries, but in fact they are (in so far as they were true) taken up into the subsequent doctrines and included in them . . . The earlier truths are not expelled but absorbed, not contradicted but extended; and the history of each science which may thus appear like a succession of revolutions, is, in reality, a series of developments.[34]

Whewell's claims must be tested against the history of developments. Consider how the velocity of light has been evaluated by physicists from Aristotle onwards. Some of the highlights of the story are presented briefly in table 1. At first sight they seem to confirm Whewell's thesis.

The realist (by which I mean one who believes that science provides truthful descriptions of nature) need not be unduly worried by the

1. Velocity of Light

Scientist	Velocity in kilometres per second	Method
Aristotle	Infinite	–
Galileo (1638)	Finite	Too great to measure
Descartes (1644)	Instantaneous	–
Romer (1676)	225,000	Observation of satellites of Jupiter
Bradley (1728)	295,000	Aberration of light
Fizeau (1849)	315,000	Rapidly turning toothed wheel.
Foucault (1862)	298,000	Rapidly rotating mirror
Michelson (1879)	299,910	Rapidly rotating mirror
(1926)	299,796	Rapidly rotating mirror
1950s	299,792.5	Electronic*
1986	299,792.4	Conversion factor between time and length: a metre is the distance light takes 1/299,792,458th of a second to travel.**

*With a reported accuracy of 1 to 3 metres.
**The velocity of light is no longer measured directly. The second is defined very accurately on the basis of the frequency of caesium atoms. The metre is then defined as the distance light of a fixed velocity travels in 1 second. One consequence of this decision, accepted by the Seventeenth General Conference of Weights and Measures in 1983, is that the speed of light is no longer open to revision. Any further refinements would leave the speed of light untouched while the length of the metre would be suitably adjusted.

fact that many areas of science do not show the same progressive development as that shown in the determination of light's velocity. As was seen earlier, there are issues such as the age, size, and density of the universe in which there seems to be very little consistent development. It is sufficient for the realist to maintain that substantial parts of science satisfy Whewell's thesis; it is not required that *every* aspect of science should conform. Given time, the realist can argue, and even issues of cosmology could begin to reveal the same kind of progressive development as that seen in other areas.

The example of the velocity of light also shows the irrelevance of theory and revolution to the realist case. Theories of the nature of light have varied considerably from the time of Romer's first measurements. Ethers have come and gone, numerous new properties of light have been discovered, and the theories of Newton, Huygens, and Young have been replaced by those of Einstein and Planck. Yet throughout all these revolutionary changes a steady and continuous progress seems to have been made in the quest for a more accurate measurement of the velocity of light.

It is, of course, possible that the progress so far witnessed is spurious. We may find in the future that scientists' estimations of light's velocity begin to vary in a more haphazard manner. After a period in which the velocity is held to be only a few thousand miles per hour, figures of millions or even billions of miles per second might be favoured.

One response to the question of whether science does make progress has been to identify specific areas of science where realism holds sway, and other areas where it is less evident. Ian Hacking, for example, has argued recently that there are two kinds of scientific realism, one for theories and one for entities. Thus, while we may well lack a satisfactory theory about the behaviour of electrons, we can still accept that electrons exist. Hacking is therefore prepared to argue that 'Protons, photons, fields of force, and black holes are as real as toe-nails, turbines, eddies in a stream, and volcanoes.'[35] But, when we come to theories, even though they may be 'constantly revised', and even though 'we use different and incompatible models of electrons which one does not think are literally true', we still accept that 'there are electrons, nonetheless.'[36] A similar distinction has been drawn by Rom Harré: like Hacking, he rejects 'truth realism', while accepting 'referential realism'.[37]

But why should the distinction be accepted without question? Why, if at all, should we be so confident of the existence of such entities as electrons, protons, and photons? The drawing of a distinction carries no existential guarantees. Realists in defence of their position normally at this point turn to the common cause argument.

The common cause argument

The argument notes that scientists often reach the same conclusion from several different and independent directions. Avogadro's

2. Values of Avogadro's constant

Scientist	Value of constant (N)	Method
Maxwell (1873)	4.3×10^{23}	Kinetic theory
Planck (1897)	6.1×10^{23}	Black body radiation
Kelvin (1904)	5.5×10^{23}	Scattering of light
Rutherford and Geiger (1908)	6.2×10^{23}	Alpha-particle emission
Perrin (1908)	6.5×10^{23}	Brownian motion

number, for example, refers to the number of molecules in one mole of any pure substance. It has a value of 6.023×10^{23}, a precise and large number unlikely to be stumbled upon by accident. Attempts were first made to calculate the constant by Loschmidt in 1865, using assumptions derived from the kinetic theory of gases.[38] He was followed over the next half-century by several other scientists who, using a variety of independent approaches, established a fairly uniform value. Some of these results, and their basis, are listed in table 2.

A fuller list, published in Perrin's *Les atomes* (1914), contained thirteen independent calculations of N. They varied between a low of 6.0×10^{23} to a high of 7.5×10^{23}, and prompted Perrin to enthuse about 'the miracle of such precise agreements coming from phenomena so different'.[39]

Results of this type have been seized upon by philosophers as conclusive proof of realism. Wesley Salmon, for example, has argued that agreement between calculations based upon Brownian motion, alpha decay, X-ray diffraction, and electrochemistry, clearly shows that 'such micro-entities as atoms, molecules, and ions' really do exist. If they did not exist, 'the striking numerical agreement in their results would constitute an utterly astonishing coincidence.' In the same way, Salmon added, we would find it extremely difficult to reject the testimony of five independent witnesses that they had seen John Doe at the scene of the crime.[40]

Conversely, in the absence of common causes, we may feel reluctant to accept a particular claim. Sometimes the evidence for the existence of a particle, a field, a force comes from just one source, using just one technique; while held to be interesting, this is seldom taken to be conclusive proof. If effects are genuine, and if photons and electrons really do exist, then they should be met with in a number of contexts.

Anti-realism

A strong attack on this argument has come from Laudan.[41] While ignoring the specific argument as applied to Avogadro's constant, he attributes to realists the principle that 'If a theory is successful we can reasonably infer that its central terms refer.' But, Laudan argues, if we examine the successful theories of eighteenth-century chemistry and physics, we find a reliance on several imponderables, subtle fluids, and ethers. As we no longer accept the existence of the ether, caloric, or phlogiston, it appears that successful theories need not refer. Equally successful, for over 2,000 years, numerous humoral theories in medicine and physiology considered disease and health to be functions of the proportions of blood, bile, black bile, and phlegm found in the body. As understood by the Hippocratics and their successors, the four humours are as fictional as caloric and phlogiston.

The response of the empiricist van Fraasen has also been critical. It is not necessary, he argues, to assume that science gives us true theories of what the world is really like; rather science 'aims to give us theories which are empirically adequate'.[42] Experience, he insists, in the classical empirical tradition, can give us information only about what is *observable* and the *actual*. And, of course, molecules are not observable. Theories about molecules, however, can be adequate by allowing us to derive truths about what is observable, namely, the manner in which volumes of different gases combine.

The sociologists' contribution

Following the example of Robert Merton in the 1930s, a number of sociologists began to look with an extremely critical eye at the actual workings of science.[43] They had all read Kuhn and Evans-Pritchard and, unlike an earlier generation of sociologists, were familiar with the work of historians and philosophers of science. Many were led to a form of relativism and a rejection of rationalism in science. Thus,

Barnes and Bloor have put forward the postulate that 'all beliefs are on a par with one another with respect to the causes of their credibility.'[44] This view they term 'monism', and they contrast it with the dualism which insists that there is a crucial distinction between true and false, rational and irrational beliefs. The dualist feels strongly that the Azande farmer's claim that his crops were spoilt by witchcraft is in need of explanation. How could anyone come to hold such an absurd belief, he asks, and how could he persist in his error? No similar feeling is held about the western farmer who explains his crop failure by an attack of blight. It is sufficient to be told that the explanation is true. In the same way, the dualist maintains, it makes little sense to ask why someone insists on believing that $5 \times 4 = 20$; if, however, we met someone who consistently maintained that $5 \times 4 = 18$, we would search for the cause of his error.

The facts, the dualist maintains, speak for themselves. The existence of electrons is sufficient to account for the physicist's belief in them; a belief, however, in the existence of fairies will call for a quite different kind of explanation. Barnes and Bloor offer a number of arguments in reply. They first point to a growing number of case studies designed to show that the facts do not speak for themselves. Consider, for example, the case of Priestley and Lavoisier in the late eighteenth century, as they observed the combustion of various substances. Although they performed virtually the same experiments, Priestley thought he had demonstrated the existence of phlogiston, while Lavoisier insisted that oxygen had been produced. Hence, Barnes and Bloor claim that 'the effect of the "facts" is neither simple nor sufficient to explain what needs explaining, viz. the theoretical divergence.'[45] Both Priestley and Lavoisier were in possession of the same set of facts; their judgements, however, were markedly different.

Rather than pursuing the details of ancient chemical experiments, the dualist may at this point concede the example to the relativist. He may also allow that many similar case histories could be constructed. At the same time, he will insist that in science and elsewhere there exists a central core of true beliefs which defy the sociologist's analysis. There are cultural universals, flouted at the cost of descending into incoherence. Thus Martin Hollis has claimed that in the absence of certain laws of logic there simply could not be any kind of reasoning at all. For Hollis, the laws $p \lor -p$, $p \And -p$, $[(p \And (p \to q)) \to q]$ express 'more than axioms in a particular system'. Rather, they express 'requirements for something's being a system of logical reasoning at all'.[46] Other communities may well have alternative systems of medicine and agriculture; they cannot also have an alternative logic. As well suppose that they could have an alternative arithmetic, in which $2 + 2 = 5$.

The point is denied by Barnes and Bloor. Logic, they insist, 'is admirable material for sociological investigation and explanation'. The supposedly compelling character of logic, they claim, derives not from reason but 'from certain narrowly defined purposes and from custom and institutionalised usage'.[47] What then, of the laws referred to by Hollis? They can hardly be considered to have universal dominion when, as is well known, intuitionist logicians reject the law of excluded middle; and in the logic of entailment developed by Anderson and Belnap, *modus ponens* is also discarded.[48] Consequently, they conclude, 'The rationalist goal of producing pieces of knowledge that are both universal in their credibility *and* justified in context-independent terms is unattainable.'[49]

Yet Barnes and Bloor seem to have overlooked one such principle. And it is indeed a principle they themselves frequently adopt. It is also a principle that is found in all cultures, at all times, and one impossible to reject on rational grounds. The principle is, in short, one of the universals of reason. It can be seen in use against the claim of Hollis that *modus ponens* is a universal requirement of reason. Not so, say Barnes and Bloor, as the principle has been rejected by a number of logicians. Their judgement at this point is a straightforward use of the Principle of Counterexample. The principle simply states:

> No universal principle or proposition against which there are agreed counterexamples need be accepted as true.

Barnes and Bloor tacitly accept the truth of this principle in their rejection of the claims of Hollis and, indeed, at numerous other places. And, of course, if they were to find the principle unacceptable then they would no longer be able to reject Hollis's espousal of the universality of *modus ponens*. Either way they seem committed to recognising that there is at least one principle of universal rationality.

Of course, in reality, there is likely to be endless debate about whether particular cases are genuine counterexamples. But the fact that we argue about whether or not x is really an F, does not mean that anyone could ever, or would ever, argue both that x is not an F, but it is true that 'all xs are F.' That is, people may well convince themselves that, although a biped and feathered, the cassowary is not in fact a bird. Having committed themselves to this view they cannot, however, also maintain that all feathered bipeds are birds. The validity of the Principle of Counterexample clearly remains unchallenged by the example.

The sociological case has, also, been overstated in other areas. It is well to be reminded that at certain times, and in certain places, the 'facts' of various matters have been allowed to take second place to personal, institutional and even political values. The 'Aryan physics' developed by Nazi scientists in the 1930s, and much of the early history of IQ testing, were both clearly designed to serve political rather than scientific ends.[50] Both sorry tales, out of which few come with any credit, have been well documented. Again, it must be conceded that scientists often bring with them to their work, along with a variety of cultural and intellectual presuppositions, excessive amounts of personal ambition. Their supposedly objective judgements can be, and inevitably sometimes are, influenced by such extraneous factors.

Yet, it can be objected, however prevalent and intrusive factors of this kind may be, they cannot ever be the whole story. For one thing scientists sometimes become aware of their own limitations and, in a moment of illumination, come to see how they have been misled by quite irrelevant factors. Typical of this approach would be the reaction of some of the leading Roman astronomer-priests to the work of Galileo. Their first response when they heard that Galileo had begun to overthrow the tradition of centuries was one of scorn. Not only did his claims that Jupiter had moons and that the moon had mountains conflict with Aristotle, they also seemed theologically suspect. It is hardly surprising that Galileo's first report in 1610 excited more criticism than admiration. Yet within a remarkably short time the same astronomer-priests were queuing up to express their agreement with Galileo. The point was expressed by Christopher Grienberger, a Jesuit mathematician, in 1611: 'I know how difficult it is to dismiss opinions sustained for many centuries by the authority of so many scholars. And surely, if I had not seen, so far as the instruments allowed, these wonders with my own eyes . . . I do not know whether I would have consented to your arguments.'[51] That is, however much a victim of his past and background a scientist may be, he can often overcome these constraints, once exposed to clear and definitive arguments and observations.

Further, and finally, the extraneous factors which so often influence a scientist's judgements are likely to be known to his colleagues, and will certainly be apparent to later historians. The historian, therefore, will often be in a position to distinguish between those aspects of a scientist's career guided by rational factors, and those controlled by other and more personal values. The point is that the historian can actually make such a distinction, and separate within a man's opus the rational and the personal. Yet, despite the intense scrutiny in the last twenty years or so of the lives and personalities of Newton and Darwin, it is surprising how little impact this work has had upon our understanding of the *texts* of *Principia* and *The Origin of Species*. We certainly know much more about the sources of Newton's ideas, how long he had held them, and what he hoped to do with them, than did scholars of a generation ago. None of this, however, can be of any relevance when the reader approaches, for example, Proposition 26 of Book 2 of *Principia*, and begins to consider the claim that: 'Pendulous bodies, that are resisted in the ratio of the velocity, have their oscillations in a cycloid isochronal.' At this point history, biography, sociology, and even philosophy, become mere distractions, as arguments are being advanced, and conclusions established.

NOTES

Introduction

1. 'It may be salutary to place a moratorium on discussion of the state, or virtue, or the moral law, and consider instead . . . the difference between kindness and kindliness' (G. J. Warnock, *Listener*, 7 April 1960, p. 617). And from J. L. Austin: 'How much it is to be wished that similar field work will soon be undertaken in aesthetics; if only we could forget for a while about the beautiful and get down instead to the dainty and the dumpy' (*Philosophical Papers*, Oxford, 1970, p. 183).

2. *The Art of the Soluble*, Harmondsworth, 1969, p. 169n. The theoretical physicist Richard Feynman claims: 'In the early fifties I suffered temporarily from a disease of middle age: I used to give philosophical talks about science' (*Surely You're Joking, Mr Feynman*, London, 1986, p. 279).

3. *The Philosophy of the Inductive Sciences*, 1840, vol. 1, p. 113.

4. *The Correspondence of Isaac Newton*, ed. H. W. Turnbull, vol. 2, Cambridge, 1960, p. 300–3.

5. W. Heisenberg, *Physics and Philosophy*, London, 1959; D. Bohm, *Causality and Chance in Modern Physics*, London, 1957; ibid., *Wholeness and the Implicate Order*, London, 1980; P. Medawar, *Induction and Hypothesis in Scientific Thought*, London, 1969.

6. Karl von Baer (1792–1876) is well known to historians of science for his fundamental discovery in 1828 of the mammalian egg. He also tackled the still largely unsolved problem of development. Working mainly with chicken embryos, von Baer claimed to have detected a very general principle at work. Development, he insisted, was always from the homogeneous to the heterogeneous. Before, for example, tissue could be identified as a mammalian heart, it could first be seen as a vertebrate heart.

7. Spencer's work is apparently little read by scientists today. When Medawar, in preparation for his 1963 Spencer lecture, consulted the Royal Society's copy of Spencer's *Principles of Biology* (2 vols, 1864–7), he found that it had been untouched for fifty years.

8. H. Spencer, *Autobiography*, London, 1904, vol. 1, p. 384.

9. Popper has recalled with horror a discussion with Adler in which he reported to him a case he knew of a disturbed child. Without having seen the child Adler immediately and confidently diagnosed an inferiority complex. Popper continued: 'Slightly shocked, I asked him how he could be so sure. "Because of my thousand fold experience," he replied; whereupon I could not help myself saying: "And with this new case, I suppose, your experience has become thousand-and-one-fold"' (*Conjectures and Refutations*, London, 1963, p. 35).

10. K. R. Popper, *Unended Quest*, London, 1976, p. 38.

11. P. A. Schilpp, *The Philosophy of G. E. Moore*, La Salle, Illinois, 1942, p. 14.

12. 'For it is a characteristic position of a positivist to hold that all symbols, other than logical constants, must either themselves stand for sense-contents or else be explicitly definable in terms of symbols which stand for sense-contents. It is plain that such physical symbols as "atom" or "molecule" or "electron" fail to satisfy this condition, and some positivists ... have been prepared to regard the use of them as illegitimate' (A. J. Ayer, *Language, Truth, and Logic*, London, 1946, p. 136).

13. 'Natural laws do not have the character of propositions which are true or false, but rather set forth instructions for the formation of such propositions ... Natural laws are not universal implications, because they cannot be verified for all cases; they are rather directions, rules of behaviour, for the investigator to find his way about in reality, to anticipate certain events' (M. Schlick in 1931, cited in W. Kneale, *Probability and Induction*, Oxford, 1949, p. 76).

14. See, for example, R. Swinburne, *An Introduction to Confirmation Theory*, London, 1973.

15. See M. J. Mulkay, *The Social Process of Innovation*, London, 1972, and B. Barnes, *Scientific Knowledge and Sociological Theory*, London, 1974.

1 When did Philosophy and Science Diverge?

1. A. J. Ayer, *Metaphysics and Common Sense*, London, 1967, p. 3.

2. R. Rorty, *Philosophy and the Mirror of Nature*, Oxford, 1980, p. 131.

3. *The Correspondence of Isaac Newton*, ed. H. W. Turnbull, vol. 3, Cambridge, 1961, pp. 71–7.

4. G. S. Kirk and J. E. Raven, *The Presocratic Philosophers*, Cambridge, 1960, p. 228.

5. W. K. C. Guthrie, *Socrates*, Cambridge, 1971, p. 98.

6. G. E. R. Lloyd, *Magic, Reason, and Experience*, Cambridge, 1979, p. 139.

7. Aristotle, *On the Parts of Animals*, 642.a.28.

8. Cicero, *Tusculan Disputations*, 5.4.10.

9. Xenophon, *Memoirs of Socrates*, Harmondsworth, 1970, pp. 30–1, 222–3.

10. Plato, *Phaedrus*, 229.e.

11. Aristotle, *Metaphysics*, 1025.b.25.

12. For Aristotle logic had no special place within his classification. As a tool, or *organon*, it was applicable to all disciplines.

13. W. and M. Kneale, *The Development of Logic*, Oxford, 1962, p. 139.

14. It even found its way into Islam. Seyyed Hossein Nasr, for example, in his *Islamic Science* (London, 1975, pp. 13–17), presents a number of classifications of the sciences which are clearly based on Stoic and Aristotelian distinctions.

15. Further details of medieval classifications can be found in J. A. Weisheipl, 'The Nature, Scope, and Classification of the Sciences', in D. C. Lindberg (ed.), *Science in the Middle Ages*, Chicago, 1978, pp. 461–82. More modern systems are described in R. Darnton, 'Philosophers Trim the Tree of Knowledge', in his *The Great Cat Massacre*, Harmondsworth, 1985, pp. 185–7.

2 How do Philosophy and Science Differ?

1. R. Scruton, *A Short History of Western Philosophy*, London, 1984, pp. 4–5.

2. G. G. Simpson, *The Major Features of Evolution*, New York, 1942, pp. 252–4.

3. The point can be followed in greater detail in J. Needham, 'Human Law and the Laws of Nature', in his *The Grand Titration*, London, 1969, pp. 299–330.

4. G. P. Thomson, *J. J. Thomson*, London, 1964, p. 114.

5. Some theoretical physicists have even been known to prefer the barstool to the laboratory bench. Feynman would make frequent visits to his local topless bar in Pasadena, where he would 'sit in one of the booths and work a little physics on the paper placemats' (*Surely You're Joking, Mr Feynman*, London, 1986, p. 270). Leo Szilard would settle in a bath in the Strand Palace Hotel at nine in the morning and 'just soak there and think' until disturbed by a maid at noon (S. Weart and G. Szilard (eds), *Leo Szilard*, London, 1978, pp. 19–20). During such a reverie

Szilard was led in 1934 to see the possibility of a nuclear chain reaction.

6. For the link between Malthus and natural selection see Antony Flew, *Darwinian Evolution*, London, 1984, pp. 74–83.

7. A. Wallace, *My Life*, London, 1908, pp. 189–90.

8. C. Darwin and T. H. Huxley, *Autobiographies*, Oxford, 1974, p. 71.

9. Kekulé was travelling on a London omnibus to Clapham pondering like a good organic chemist upon the structure of carbon: 'And lo, the atoms were gamboling before my eyes ... I saw frequently how two smaller atoms united to form a pair; how a larger one embraced two smaller ones ... I saw how the longer ones formed a chain ... [and then] the cry of the conductor "Clapham Road" awakened me from my dreaming' (D. L. Hurd and J. J. Kipling, *The Origins and Growth of Physical Science*, Harmondsworth, 1964, vol. 2, p. 124).

10. D. M. Raup, *The Nemesis Affair*, New York, 1985, p. 112.

11. R. Wollheim (ed.), *Hume on Religion*, London, 1963, p. 38.

12. Ibid., p. 39n.

13. Ibid., p. 45.

14. On the accuracy of Linnaeus see p. 239 below.

15. B. Russell, *Mysticism and Logic*, Harmondsworth, 1953, p. 106.

16. A. J. Ayer, *Metaphysics and Common Sense*, London, 1979, pp. 87–8.

17. Aristotle, *Metaphysics*, ed. J. Warrington, London, 1956, p. 115.

18. J. Ree, M. Ayers, and A. Westoby, *Philosophy and its Past*, Brighton, 1978, p. 1.

19. R. Rorty, J. B. Schneewind, and Q. Skinner, *Philosophy in History*, Cambridge, 1984, pp. 49–50.

20. D. J. de Sola Price, *Little Science, Big Science*, New York, 1986.

21. J. C. Eccles, *The Human Mystery*, London, 1984, lecture 10, *passim*.

22. P. Davies, *God and the New Physics*, Harmondsworth, 1984, pp. 186–7. For a fuller account of the anthropic principle J. Barrow and F. Tipler, *The Anthropic Cosmological Principle*, Oxford, 1986 should be consulted.

23. See chapter 7 below.

24. J. L. Austin, *Philosophical Papers*, Oxford, 1970, pp. 182–3.

3 Is Science Really the Art of the Soluble?

1. John Ziman, *Public Knowledge*, Cambridge, 1968, p. 9.

2. P. Medawar, *The Art of the Soluble*, Harmondsworth, 1969, p. 97.

3. G. Ryle, 'Autobiographical', in O. P. Wood and G. Pitcher (eds), *Ryle*, London, 1971, p. 12.

4. J. L. Austin, *Philosophical Papers*, Oxford, 1970, p. 175.

5. H. F. Judson, *The Eighth Day of Creation*, London, 1979, p. 496.

6. Thus, Gerald Holton has noted that Bohr 'tended to regard Fermi's

solutions as too simple to be profoundly important'. Further, 'Bohr really had it in his mind that there was some profound problem with neutrinos and energy . . . and did not want to have it solved except in a mystical and deep way. It was solved by Fermi in "too elementary" a way' (*Striking Gold in Science*, Minerva, 1974, p. 165).

7. Chargaff had established in 1948 the vital base ratios of DNA, namely, guanine/cytosine = adenine/thymine = 1. It was these ratios which later led Watson and Crick to the structure of DNA. Chargaff has since commented on the results of his work: 'My main objection to molecular biology is that by its claim to be able to explain everything it actually hinders the free flow of scientific ideas but there is not a scientist I have met who would share my opinion' (Judson, *Eighth Day*, p. 222).

8. B. Hoffmann, *Albert Einstein*, London, 1973, p. 228.

9. R. W. Clark, *Einstein: The Life and Times*, London, 1979, p. 384.

10. H. M. Edwards, 'Fermat's Last Theorem', *Scientific American*, October 1978, pp. 86–97.

11. As, on one memorable occasion in his world championship match against Spassky, Bobby Fischer played BxKRP and thereby gave Spassky a bishop. 'Unbelievable', commented the analyst Alexander. And Fischer's own comment was: 'I played like a fish.' See C. H. O'D. Alexander, *Fischer v. Spassky*, Harmondsworth, 1972, pp. 85–7.

12. Walter Korn (ed.), *Modern Chess Openings*, 11th edition, London, 1972.

13. David Armstrong, *Belief, Truth and Knowledge*, Cambridge, 1973, pp. 152–61.

14. The critic can also be a figure to be feared and avoided. Watson, for example, has spoken of the Cambridge reputation established by the Francis Crick of pre-DNA days: 'There existed an unspoken yet real fear of Crick, especially among his contemporaries who had yet to establish their reputations. The quick manner in which he seized upon their facts and tried to reduce them to coherent patterns frequently made his friends' stomachs sink with apprehension that, all too often in the near future, he would succeed, and expose to the world the fuzziness of minds hidden from direct view by the considerate, well-spoken manners of the Cambridge colleges' (*The Double Helix*, London, 1968, p. 10).

15. Peter Vorzimmer, *Charles Darwin: The Years of Controversy*, London, 1972, pp. 148–54.

16. The story of Von Pettenkofer can be followed in C.-E. Amory Winslow, *The Conquest of Epidemic Disease*, Wisconsin, 1971, chapter 15.

17. In the same vein a modern critic has challenged the orthodox judgement that AIDS is transmitted by HIV. The critic, Peter Duesberg, a molecular biologist from Berkeley, is presumably as knowledgeable

about viruses as Von Pettenkofer was about bacteria. Also, in a manner reminiscent of Von Pettenkofer, Duesberg has announced his willingness to inject himself with pure HIV. See P. Duesberg, 'AIDS and the "Innocent" virus', *New Scientist*, 28 April 1988, pp. 34–5, and the reply by Jonathan Weber, 'AIDS and the "guilty" virus', *New Scientist*, 5 May 1988, pp. 32–3.

18. Their arguments can be followed in their works: *Lifecloud*, London, 1978; *Diseases from Space*, London, 1979; *Space Travellers*, London, 1981; and *Archaeopteryx, the Primordial Bird: a Case of Fossil Forgery*, London, 1986.

19. Lord Macaulay, *Critical and Historical Essays*, London, 1849, vol. 2, p. 399.

20. Seneca, *Letters from a Stoic*, ed. R. Campbell, Harmondsworth, 1969, p. 172.

21. Beatrice Webb, *My Apprenticeship*, Harmondsworth, 1938, p. 153.

22. Nick Herbert, 'How to be in Two Places at One Time', *New Scientist*, 21 August 1986, p. 41.

23. P. Davies and J. Brown (eds), *The Ghost in the Atom*, Cambridge, 1986, p. 60.

24. T. Hey and P. Walters, *The Quantum Universe*, Cambridge, 1987, p. 127.

25. P. Medawar, *The Art of the Soluble*, Harmondsworth, 1969, p. 97.

26. Even more than Fermat's last theorem, Riemann's hypothesis is the great unsolved problem of modern mathematics. It is discussed in K. Devlin, *Mathematics: The New Golden Age*, Harmondsworth, 1988, pp. 211–21. How would a philosopher respond to the question? Ayer has described how he once teased the Cambridge philosopher A. C. Ewing by asking him 'what he most looked forward to in the next world. He replied without hesitation "God will tell me whether there are synthetic a priori propositions"' (A. J. Ayer, *More of my Life*, London, 1985, p. 15).

27. The bulk of the figures are taken from A. van Helden, *Measuring the Universe*, Chicago, 1985, *passim*.

28. S. J. O'Brien, 'The Ancestry of the Giant Panda', *Scientific American*, November 1987, pp. 82–7.

29. Referred to in A. Kohn, *False Prophets*, Oxford, 1986, p. 8.

30. A. P. French (ed.), *Einstein*, London, 1979, p. 23.

31. David Armstrong, 'The Nature of Mind', in C. V. Borst (ed.), *The Mind/Brain Identity Theory*, London, 1970, p. 69.

32. Robert Latham (ed.), *The Diaries of Pepys*, London, 1970, vol. 1, pp. lxxix–lxxxii.

33. L. Laudan, *Science and Values*, Berkeley, Cal., 1984, p. 13.

34. J. Kekes, *The Nature of Philosophy*, Oxford, 1980, p. 5.

35. 'What we do may be small, but it has a certain character of permanence;

and to have produced anything of the slightest permanent interest, whether it be a copy of verses or a geometrical theorem, is to have done something utterly beyond the powers of the vast majority of men' (G. H. Hardy, *A Mathematician's Apology*, Cambridge, 1967, p. 76).

36. A disjunction of the form (p v q) is true if and only if at least one of the disjuncts is true.

37. The argument can be followed in A. N. Prior, *Formal Logic*, Oxford, 1962, pp. 240–50.

38. H. Reichenbach, *Philosophical Foundations of Quantum Mechanics*, Berkeley, 1944.

39. Another possible source of error could be Medawar's long-standing role of research director. 'Research', Medawar proclaimed, not 'science', was the art of the soluble. The view makes very good sense to busy and ambitious directors of research. If students are assigned problems requiring more than a few years' work, funds and students will rapidly disappear from the laboratory. A good director is, therefore, someone sufficiently inventive to have at his disposal problems hard enough to stretch his students, yet capable of solution by the bright and industrious within the appropriate time. Once this implicit contract is violated and the doctoral student finds himself three years later without funds, job, and anything publishable, then he may well feel that it is his director's incompetence, and not his own, that has been displayed. Yet, surely, there is more to science than a successful Ph.D. dissertation!

4 Do Scientific Problems Begin in Philosophy?

1. R. F. Harrod, *J. M. Keynes*, Harmondsworth, 1972, pp. 372–4.

2. See, for example, B. d'Espagnat, *In Search of Reality*, Berlin, 1983; N. Herbert, *Quantum Reality*, London, 1985; A. Shimony, 'The Reality of the Quantum World', *Scientific American*, January 1988.

3. J. Needham, *Science and Civilisation in China*, vol. 3, Cambridge, 1959, p. 112.

4. S. Carnot, *Reflections on the Motive Power of Fire*, ed. E. Mendoza, New York, 1960, p. 3.

5. Ibid., p. 5.

6. C. Darwin and T. H. Huxley, *Autobiographies*, Oxford, 1974, p. 70.

7. In 1814 Fraunhofer reported that the solar spectrum contains numerous dark lines. But the spectra of elements heated in the laboratory are marked by bright emission lines. Once a correspondence had been worked out between the dark solar lines and the bright laboratory lines, Kirchoff was able to show in the 1860s the presence in the sun of a number of common terrestrial elements.

8. *The Philosophical Writings of Descartes*, ed. J. Cottingham, R. Stoothoff, and D. Murdoch, Cambridge, 1985, vol. 1, p. 353.
9. D. Hume, *Treatise on Human Nature*, Book 2, part 1, section 5.
10. George Miller, *Psychology*, Harmondsworth, 1966, p. 111.
11. Ibid., pp. 108–13.
12. E. Tylor, *Primitive Culture*, London, 1913, vol. 1, p. 2.
13. Ibid., p. 8.
14. Ibid., pp. 32–3.
15. D. Forde (ed.), *African Worlds*, Oxford, 1970, p. 166.
16. O. Neugebauer, *The Exact Sciences in Antiquity*, New York, 1969, pp. 105, 110, 129.
17. Ibid., pp. 126–7.
18. Needham, *Science and Civilisation*, p. 399.
19. Ibid., p. 193.
20. Ibid., pp. 191–2.
21. The main ideas of Aristotelian physics can be followed in G. E. R. Lloyd, *Early Greek Science: Thales to Aristotle*, London, 1970, and I. B. Cohen, *The Birth of New Physics*, Harmondsworth, 1987.
22. See S. E. Toulmin and J. Goodfield, *The Fabric of the Heavens*, Harmondsworth, 1963, p. 129.
23. *The Correspondence of Isaac Newton*, ed. H. W. Turnbull, vol. 3, Cambridge, 1962, pp. 359–60.
24. According to Elic Howe, *The Magicians of the Golden Dawn*, London, 1972, pp. 50–1, William Crookes, President of the Royal Society from 1913 to 1915, 'took the Neophyte grade in Isis-Urania in June 1890'.
25. J. Borges, *Labyrinths*, Harmondsworth, 1970, pp. 87–95.

5 Do Philosophical Problems End in Science?

1. M. McCloskey, 'Intuitive physics', *Scientific American*, April 1983, pp. 114–22.
2. See A. G. Debus, *The English Paracelsians*, London, 1986, pp. 30–1.
3. *The Philosophical Writings of Descartes*, ed. J. Cottingham, R. Stoothoff, and D. Murdoch, Cambridge, 1985, vol. 1, p. 140.
4. The debate provoked by Descartes on this issue is discussed in N. Chomsky, *Cartesian Linguistics*, New York, 1966.
5. D. L. Hurd and J. J. Kipling, *The Origin and Growth of Modern Science*, Harmondsworth, 1964, vol. 2, p. 107.
6. J. Needham, *Man a Machine*, London, 1927, p. 31.
7. Ibid., p. 34.
8. Anatomists have long sought a feature to distinguish man from the rest of nature. A bitter dispute, for example, was waged between Robert

Owen and T. H. Huxley in the nineteenth century as to whether the hippocampus could be found in the brains of both apes and humans. But by 1903 the immunologist, Elie Metchnikoff, could conclude that the human brain contained no parts absent from the simian brain. The reproductive system, he claimed, was a better source of differentia. Man, he pointed out, lacks the *os penis*, and the hymen is peculiar to the human race (*The Nature of Man*, London, 1906, pp. 81–2).

9. See O. Chadwick, *The Secularization of the European Mind in the Nineteenth Century*, Cambridge, 1975, p. 166.

10. J. J. C. Smart, 'Sensations and Brain Processes', in C. V. Borst (ed.), *The Mind/Brain Identity Theory*, London, 1970, pp. 52–66.

11. J. C. Eccles, *The Human Mystery*, London, 1984, p. 214.

12. Ibid., p. 222.

13. Ibid., p. 224.

14. Ibid., p. 226.

15. Ibid., p. 227.

16. Further details of these and other computer languages can be found in David Waltz, 'Artificial Intelligence', *Scientific American*, October 1982, pp. 101–22.

17. Racter is described in A. K. Dewdney, 'Computer Recreations', *Scientific American*, January 1985, pp. 10–13. More of Racter's conversation can be found in W. Chamberlain's *The Policeman's Beard is Half-Constructed*, New York, 1984.

18. Waltz, 'Artificial Intelligence', pp. 120–2.

19. Searle's argument was originally contained in his 'Minds, Brains and Programs', *Behavioural and Brain Sciences*, September 1980. A more accessible version can be found in his 1984 Reith lectures: *Minds, Brains and Science*, London, 1984, chapter 2.

20. Searle, *Minds, Brains and Science*, p. 31.

21. Ibid., p. 33.

22. *The Republic*, 597.

23. *Principles of Human Knowledge*, section 6.

24. J. Gribbin, *In Search of Schrodinger's Cat*, London, 1985; N. Herbert, *Quantum Reality*, London, 1985.

25. *A Treatise on Human Nature*, Book 1, part 3, section 2.

26. Ibid.

27. J. C. Polkinghorne, *The Quantum World*, London, 1984, p. 73.

28. B. d'Espagnat, 'The Quantum Theory and Reality', *Scientific American*, November 1979, p. 128.

29. For the work of Bell and Aspect see N. Herbert, *Quantum Reality*, London, 1985; A. Shimony, 'The Reality of the Quantum World', *Scientific American*, January 1988.

30. d'Espagnat, 'The Quantum Theory', 1979, p. 128.

31. P. Davies and J. Brown (eds), *The Ghost in the Atom*, Cambridge, 1986, p. 48.

32. Herbert, *Quantum Reality*, p. 245.

33. Ibid., p. 246.

34. Ibid., pp. 214-15.

35. Shimony, 'The Reality of the Quantum World', p. 36.

36. The floor of his office, Eddington declared, had no 'solidity of substance'. To step on it was like stepping on 'a swarm of flies'. It was easier, he concluded, for a camel to pass through the eye of a needle 'than for a scientific man to pass through a door' (*The Nature of the Physical World*, Cambridge, 1932, p. 342).

6 Is there a Scientific Method?

1. Consider the sheer intellectual difficulty facing people unfamiliar with modern technology when they first witnessed air-drops. As bags of rice, flour, axes, and other goods fell from the sky they, according to one witness, 'sat in awe and wonderment'. In response, 'Large imitation US camps were set up all round Madang on deserted Army camp-sites, filled with devotees white-hot with enthusiasm At daybreak, the people "fell in" with their "commanders", "paymasters", "servants", "RT operators", "guards" and even their women and children' (P. Worsley, *The Trumpet Shall Sound*, London, 1970, p. 224).

2. On Lull, see Martin Gardner's informative essay in his *Logic Machines*, New York, 1968; Dee's *Monas* is discussed in P. J. French, *John Dee*, London, 1972; for Leibniz, see W. and M. Kneale, *The Development of Logic*, Oxford, 1962.

3. S. II. Mellone, *An Introductory Text Book of Logic*, Edinburgh, 1909, p. 247.

4. A. Lavoisier, *Elements of Chemistry*, New York, 1965, pp. 61-5.

5. Ibid., p. 71.

6. See H. Hartley, *Humphry Davy*, London, 1966, pp. 74-80.

7. F. Bacon, *The Advancement of Learning*, Oxford, 1974, p. 120.

8. Kelvin's estimates of the age of the earth's habitability actually varied considerably, from the 10 million of 1854 to his final figure in 1897 of 40 million years.

9. P. G. Tait, *Recent Advances in Physical Science*, London, 1876, p. 167.

10. D. L. Eicher, *Geologic Time*, London, 1973, pp. 17-19.

11. Medieval accounts of scientific method are discussed in A. C. Crombie, *Augustine to Galileo*, London, 1961, vol. 2, pp. 11-25.

12. Book 1, part 3, section 15.

13. Book 3, chapter 8.

14. Snow found that all those infected with cholera in the 1845 Soho epidemic had drunk water drawn from a pump in Broad Street. The pump was closed and the disease rapidly disappeared from Soho. First described in 1981, AIDS was identified as a sexually transmitted disease when such symptoms as Kaposi's sarcoma, known mainly to infect older Jewish and Italian men, began to turn up untypically in younger homosexual men.

15. *Of Induction*, London, 1849, p. 44.

16. This line of argument is developed in considerable detail in H. L. A. Hart and A. Honoré, *Causation in the Law*, Oxford, 1959.

17. 'Studies in the Logic of Confirmation', *Mind*, 1945, pp. 1–26, 97–121. They are also included in Hempel's *Aspects of Scientific Explanation*, New York, 1965, pp. 3–51.

18. *Fact, Fiction, and Forecast*, London, 1955.

19. M. Gardner, 'On the Fabric of Inductive Logic', *Scientific American*, March 1976, pp. 119–22.

20. The story can be followed in W. and M. Kneale, *The Development of Logic*, pp. 652–88, 712–24.

21. Book 1, part 3, section 6.

22. Plato, *Timaeus*, ed. H. D. P. Lee, Harmondsworth, 1965, p. 41.

23. A. Koestler, *The Sleepwalkers*, Harmondsworth, 1964, pp. 573–4.

24. T. Kuhn, *The Copernican Revolution*, Cambridge, Mass., 1966, p. 138.

25. *The Philosophical Writings of Descartes*, ed. J. Cottingham, R. Stoothoff, and D. Murdoch, Cambridge, 1985, vol. 1, p. 289.

26. Ibid., p. 290.

27. *Works of Thomas Reid*, 1863, Edinburgh, 6th edn, vol. 1, p. 250.

28. Ibid., p. 207.

29. Ibid., p. 250.

30. See G. Cantor, *Optics after Newton*, Manchester, 1983, pp. 150–2.

31. W. Whewell, *The Philosophy of the Inductive Sciences*, London, 1840, vol. 2, pp. 62–4.

32. J. S. Mill, *System of Logic*, Book 3, chapter 14, section 6.

33. B. Hoffmann, *Einstein*, London, 1975, p. 133.

34. In his *Variation of Animals and Plants* (1868) Darwin proposed a theory of inheritance in which all the cells of the body gave off gemmules or *pangenes* which eventually moved to the sexual organs and became part of the sex cells. In this manner events affecting the body could influence the pangenes and thereby be inherited. Among other things the theory of pangenesis could serve as a basis for the controversial claim that

acquired characteristics could be inherited. The crucial experiment was performed by Francis Galton in 1871. Silver-grey rabbits were transfused with blood from the common lop-eared. Yet when the silver-grey bred, their offspring showed not a single lop-eared trait. D. W. Forrest, *Francis Galton*, London, 1974, pp. 102–9.

35. Cited by P. Vorzimmer, *Charles Darwin: The Years of Controversy*, London, 1972, p. 149.

36. Ibid., p. 152.

37. R. Feynman, *Surely You're Joking, Mr Feynman*, London, 1986, pp. 253–5.

38. See R. T. Petersson, *Sir Kenelm Digby*, Cambridge, Mass., 1956.

39. J. Golding, S. Limerick, and A. MacFarlane, *Sudden Infant Death*, Shepton Mallet, 1985, p. 8.

40. Ibid., p. 8. Further evidence comes from the surgeon Geoffrey Keynes, who could record in 1939, after twenty-five years of successful practice, that he had never 'even seen the thymus gland'. *The Gates of Memory*, Oxford, 1983, p. 276.

41. R. Hare, *The Birth of Penicillin*, London, 1970, p. 66.

42. Ibid., p. 87.

43. W. I. B. Beveridge, *The Art of Scientific Investigation*, London, 1957, pp. 167–8.

44. A. Hallam, *A Revolution in the Earth Sciences*, Oxford, 1973, p. 25.

45. Ibid., p. 24.

46. Ibid., p. 20.

47. See D. Oldroyd, *The Arch of Knowledge*, London, 1986, pp. 243–5.

7 Can Philosophy Learn from Science?

1. B. Russell, *Autobiography*, London, 1971, vol. 1, p. 151.

2. I. C. Jarvie, *The Revolution in Anthropology*, London, 1964; E. Sober, *The Nature of Selection*, Cambridge, Mass., 1985; I. Hacking, 'Microscopes', in his *Representing and Intervening*, Cambridge, 1983, pp. 186–209; H. Krips, *The Metaphysics of Quantum Theory*, Cambridge, 1987.

3. The success of the germ theory can be followed in C.-E. A. Wilson, *The Conquest of Epidemic Disease*, Madison, Wisconsin, 1980. The problem of malaria is discussed in G. Harrison, *Mosquitoes, Malaria and Man*, London, 1978.

4. For example, A. De Bary, *Lectures on Bacteria*, Oxford, 1887, pp. 170–1.

5. N. Hanson, 'The Zenith and Nadir of Newtonian Mechanics', *Isis*, 1962, pp. 359–78.

6. B. Hoffmann, *Einstein*, London, 1975, pp. 124–33.

7. The story of the survival and eventual decline of Euclid is told in my *The Classics of Science*, New York, 1984.

8. Russell, *Autobiography*, p. 36.

9. Examples of Euclid's optical postulates are (a) that rectilinear rays proceeding from the eye diverge indefinitely; (b) that the figure contained by a set of visual rays is a cone of which the vertex is the eye and the base at the surface of the objects seen.

10. W. L. Letwin, *The Origins of Scientific Economics*, London, 1963.

11. E. Kasner and J. Newman, *Mathematics and the Imagination*, Harmondsworth, 1958, pp. 181-2.

12. The proof is short enough to be given in outline. In Book XI, Theorem 21, Euclid had proved that any solid angle is contained by plane angles less than four right angles. Therefore the angles of a regular solid could only be formed from:

> 3, 4, or 5 angles of 60° from 3, 4, or 5 regular triangles,
>
> 3 angles of 90° from 3 squares,
>
> 3 angles of 108° from 3 regular pentagons.

Any other combination would yield the impossibility of solid angles greater than or equal to 360°. Thus there cannot be, as a matter of logic, any regular solids formed from any number of regular hexagons. The angles of a regular hexagon equal 120°, and as it takes at least three plane angles to form a solid angle, no combination of regular hexagons could ever form a solid angle less than 360°.

13. *The Philosophical Writings of Descartes*, ed. J. Cottingham, R. Stoothoff, and D. Murdoch, Cambridge, 1985, vol. 1, p. 120.

14. The *Meditations* was first published in 1641 in Latin. Before publication the manuscript had been shown to such leading intellectuals as Antoine Arnaud, Pierre Gassendi, Thomas Hobbes, and Marin Mersenne, and their comments invited. The *Meditations* thus appeared with six sets of 'Objections' together with Descartes's 'Replies'. In a second edition published in 1642 a seventh set of 'Objections' from the Jesuit Pierre Bourdin was added.

15. H. Midonick, *The Treasury of Mathematics*, Harmondsworth, 1968, vol. 1, pp. 401-2.

16. Russell and Whitehead proved that 1 is a number in Proposition 52.16, p. 345 of their *Principia Mathematica*, Cambridge, 1910, vol. 1.

17. Descartes, *Philosophical Writings*, vol. 2, p. 111.

18. Ibid., p. 117.

19. In his *The World*, Descartes distinguished between three elements. Parts of the first element are so small and so rapid that they cannot be stopped; those of the second element are so moderate in their motion and size that while 'there are many causes in the world which may increase their

motion and decrease their size, there are just as many others which can do exactly the opposite.' Finally, the parts of the third element are so large as always to have 'the force to resist the motions of other bodies' (Descartes, *Philosophical Writings*, vol. 1, p. 89).

20. Ibid., p. 323.

21. Spinoza, *Works*, New York, 1955, vol. 2, p. 129.

22. For example, Gottlob Frege published the second volume of his *Grundgesetze der Arithmetik* (Jena, 1903) with a postscript which began: 'Hardly anything more unfortunate can befall a scientific writer than to have one of the foundations of his edifice shaken after the work is finished.' He was referring to a letter from Bertrand Russell pointing out that it was possible to derive a contradiction from Frege's fifth axiom. Attempts by Frege and others to modify his system proved unsatisfactory. See W. and M. Kneale, *The Development of Logic*, Oxford, 1962, pp. 652–7.

23. Spinoza, *Works*, p. 83.

24. Ibid., p. 93.

25. E. J. Aiton, *Leibniz*, Bristol, 1985, pp. 25–6, 37–9.

26. J. Locke, *An Essay Concerning Human Understanding*, Book IV, chapter 3, section 18.

27. Hutcheson's deductivism is discussed in L. I. Bredvold, 'The Invention of the Ethical Calculus', in R. F. Jones (ed.), *The Seventeenth Century*, Stanford, 1951, pp. 164–80.

28. Published as *The Logical Structure of the World*, Berkeley, Cal., 1967.

29. S. E. Toulmin, *Human Understanding*, Oxford, 1972, pp. 52–65.

30. The dimensions of Newton's achievements can be followed in two important books by I. B. Cohen: *The Newtonian Revolution*, Cambridge, 1980; *The Birth of a New Physics*, Harmondsworth, 1987.

31. *Principia*, Book III, Prop. IV.

32. D. Hume, *A Treatise of Human Nature*, Book I, part I, section 4.

33. Id., *Enquiries Concerning the Human Understanding*, section 3.

34. Id., *Treatise*, Book 1, part 1, section 4.

35. See J. W. Reeves, *Body and Mind in Western Thought*, Harmondsworth, 1958, pp. 347–8.

36. John Stuart Mill, *A System of Logic*, Book VI, chapter 4.

37. A. Bain, *Mental and Moral Science*, London, 1872, pp. 85–161.

38. See L. Laudan, 'Thomas Reid and the Newtonian Turn of British Methodological Thought', in R. E. Butts and J. W. Davis (eds), *The Methodological Heritage of Newton*, Oxford, 1970.

39. This contrast is discussed at length in L. Laudan, 'The Medium and its Message', in G. N. Cantor and M. J. S. Hodge (eds), *Conceptions of Ether*, Cambridge, 1981, pp. 157–85.

40. I. Newton, *Opticks*, New York, 1979, pp. 345–6.
41. Ibid., pp. 353–4.
42. See selections from Hartley in Reeves, *Body and Mind*, pp. 346–51.
43. Laudan, 'The Medium', pp. 164–9.
44. K. R. Popper, 'Evolution and the Tree of Knowledge', in *Objective Knowledge*, Oxford, 1972, pp. 256–84; S. E. Toulmin, *Human Understanding*, Oxford, 1972, pp. 136–42.
45. T. H. Huxley, 'The Coming of Age of "The Origin of Species"', in his *Darwiniana*, London, 1893, p. 229.
46. R. Popper, *Objective Knowledge*, p. 261.
47. The experiments were originally performed by G. F. Gause and are described by P. Olinvaux, *Why Big Fierce Animals are Rare*, Harmondsworth, 1982, pp. 125–6.
48. James Fallows, 'The American Army and the M-16 Rifle', in D. MacKenzie and J. Wajcoma (eds), *The Social Shaping of Technology*, Milton Keynes, 1985, pp. 239–51.

8 Is Science Open to Philosophical Criticism?

1. St Augustine, *Confessions*, Harmondsworth, 1961, pp. 93–4.
2. M. Allott and R. H. Soper, *Matthew Arnold*, The Oxford Authors, Oxford, 1986, p. 465.
3. Further details can be found in John Passmore, *Science and its Critics*, London, 1978.
4. K. R. Popper, *The Open Society and its Enemies*, vol. 2, London, 1962, p. 27.
5. Cremonini's folly can be exaggerated. It is a far from straightforward matter to know what weight can be put upon the reliability and accuracy of new instruments. Many, on first using a telescope, have found it difficult to find the moon, let alone Jupiter's satellites. Further, if Cremonini's caution warrants scorn, what are we to make of Kepler, who accepted without hesitation all Galileo's claims long before he had even seen a telescope. 'I may perhaps appear rash', he mused uncomfortably.
6. The matter has been reviewed by Walter Kaufmann (*Hegel*, London, 1966, p. 77). Hegel's comments were directed, with some point, against the uncritical use made of Bode's law. Formulated by Johann Bode (1747–1826), the law stated that if to each member of the series 0, 3, 6, 12, 24 . . . we add 4 and divide by 10, a quite arbitrary procedure, we are left with the distance of the planets from the sun in astronomical units (the mean distance of the earth from the sun). In 1800 astronomers were faced with the following data:

Planet	Bode's predicted distance in AU	Actual distance in AU
Mercury	0.4	0.39
Venus	0.7	0.72
Earth	1.00	1.00
Mars	1.6	1.52
–	2.8	–
Jupiter	5.2	5.2
Saturn	10.0	9.54
Uranus	19.6	19.19

Conscious of the tight fit between Bode's law and planetary distances, some astronomers began to argue, *and for no other reason*, that there must be an undiscovered planet between Mars and Jupiter. With as little justification, Hegel argued that Plato in his *Timaeus* (ed. H. D. P. Lee, Harmondsworth, 1965, pp. 35–6) had also provided a series through which 'the demiurge constructed the universe.' And, in this series, 'between the fourth and fifth member there is a large interval in which one need not miss a planet.' What can be said other than that in this case the astronomers were luckier than the philosophers?

7. R. Harrod, *The Prof*, London, 1959, pp. 17–27.
8. K. R. Popper, *Unended Quest*, London, 1976, p. 93.
9. H. G. Alexander (ed.), *The Leibniz–Clarke Correspondence*, Manchester, 1956, p. 16.
10. Ibid., p. 44.
11. J. Boswell, *The Life of Dr. Johnson*, London, 1906, vol. 1, p. 292. The joke was good enough to bear repetition. On another occasion Johnson, speaking to a follower of Berkeley, marked his departure with the comment: 'Pray, Sir, don't leave us; for we may perhaps forget to think of you, and then you will cease to exist' (vol. 2, p. 334).
12. References in the rest of this section to Berkeley's works are to the Everyman edition, *A New Theory of Vision and Other Writings*, London, 1954.
13. Early theories of vision are discussed most comprehensively by D. Lindberg, *Theories of Vision from Al-Kindi to Kepler*, Chicago, 1976; much also of interest can be found in A. I. Sabra, *Theories of Light from Descartes to Newton*, Cambridge, 1981.
14. Berkeley, p. 14.
15. Ibid., p. 18.
16. Ibid., p. 55.

17. Ibid., p. 55.
18. Ibid., pp. 163–84.
19. I. Newton, *Principia*, ed. F. Cajori, Berkeley, 1962, vol. 1, pp. xvii–xviii.
20. Berkeley, pp. 164–5.
21. Newton, *Principià*, vol. 2, p. 547.
22. Berkeley, p. 173.
23. The controversy can be followed in C. B. Boyer, *The History of the Calculus*, New York, 1949, chapter 6. The strange notion that these small inconvenient terms could be simply ignored continued to be accepted in mathematical textbooks until quite recent times. Thus, in S. P. Thomson's popular text, *Calculus Made Easy* (London, 1965, pp. 15–16), the student finds that in the process of differentiating $y = x^2$, he derives the equation: $dy = 2x \cdot dx + (dx)^2$. But, Thomson intervenes, supposedly in explanation: 'What does $(dx)^2$ mean? Remember that dx means a bit – a little bit – of x. Then $(dx)^2$ will mean a little bit of a little bit of x . . . It may therefore be discarded as quite inconsiderable.' Yet nowhere is it suggested that the answer, $dy/dx = 2x$ is not an exact solution. Nor is it explained how we can ignore small quantities in the above equation, but cannot possibly accept that the following equation is correct: $2 + 2 + 0.000001 = 4$.
24. *The Analyst*, section 35.
25. K. R. Popper, 'A Note on Berkeley as Precursor of Mach and Einstein', in *Conjectures and Refutations*, London, 1969, pp. 166–74. For John Wheeler, see J. D. Barrow and F. J. Tippler, *The Anthropic Cosmological Principle*, Oxford, 1988, p. 22.
26. Feyerabend's *Science in a Free Society* (London, 1978, pp. 107–22) contains an interesting account of the genesis of some of his more startling views. After studying 'theatre, history, mathematics, physics, and astronomy', but never philosophy, he began teaching the philosophy of science at the University of Bristol in 1957. While working on quantum theory he identified a large gap between the methodological assumptions of physics and its actual practice. His great moment of illumination, however, came in the mid-1960s in Berkeley, California. Under the enlightened educational policies of the times he found himself teaching a growing number of Mexicans, blacks and Indians. Before long he came to feel that he was doing little more than alienating them from their own 'rich and complex cultures' and enslaving them to an abstract intellectualism that he could not justify. At first he thought the answer lay in reforming the tradition of western rationalism from within. But he later came to see that such fields as elementary particle physics and Kantianism 'should not be reformed, but should be allowed to die a natural death'. Another possibility, he concluded, would be to

start a career as an entertainer. Instead he seems to have adopted the role of philosophical anarchist.

27. Feyerabend, *Science*, p. 100.

28. Ibid., p. 105.

29. Feyerabend, *Against Method*, London, 1975, p. 28.

30. Feyerabend, *Science*, p. 105.

31. Friedrich Kekulé (1829–96), for example, one of the founders of modern organic chemistry, has described how the idea of the ring structure of the benzene molecule did literally first occur to him in a dream: 'One of the snakes had seized hold of its own tail, and the form whirled mockingly before my eyes. As if by a flash of lightning, I awoke and . . . spent the rest of the night working out the consequences' (D. L. Hurd and J. J. Kipling, *The Origins and Growth of Physical Science*, Harmondsworth, 1964, vol. 2, p. 104). Undoubtedly, the 'working out' would have been carried through in orthodox chemical terms.

32. *The Correspondence of Isaac Newton*, ed. H. W. Turnbull, vol. 2, Cambridge, 1960, pp. 435–40.

33. The quotation in full reads: 'Nor can I ever sufficiently admire the outstanding acumen of those who have taken hold of this opinion [Copernicanism] and accepted it as true; they have through sheer force of intellect done such violence to their own senses as to prefer what reason told them over that which sensible experience plainly showed them to the contrary' (Galileo, *Dialogue Concerning the Two Chief World Systems*, trans. S. Drake, Berkeley, Cal., 1954, p. 328). Galileo found less to admire, however, in the philosophers who insisted on preferring the reason of Aristotle to the sensible experience of Galileo's own telescope.

34. For example, to the charge that the earth's daily rotation would produce violent winds, Oresme replied: 'The daily motion affects not only the Earth, but along with it the water and the atmosphere . . . Think of the air enclosed in a moving ship: to a person in the ship, this air would appear to be stationary.' Oresme's arguments are reviewed by T. H. Kuhn in his *The Copernican Revolution*, Cambridge, Mass., 1966, pp. 114–22.

35. Thus Ptolemy in the *Almagest*: 'All who think it strange that such an immense mass as that of the Earth should neither move itself nor be carried somewhere seem to me to look to their own *personal experience*, and not to the special character of the universe, and go wrong through regarding the two things as analogous.' Ptolemy's arguments for 'The absolute immobility of the Earth' can be found in Hurd and Kipling, *Origins and Growth of Physical Science*, vol. 1, pp. 68–71.

36. See Kuhn, *The Copernican Revolution*, p. 179.

37. Copernicus, *De Revolutionibus*, Book 1, section x.
38. Such rhetorical tropes are not uncommon in science. William Harvey, for example, arguing in 1628 for the circulation of the blood, sought first to appeal to his audience by emphasising the importance of the heart: 'The heart of animals is the foundation of their life, the sovereign of everything with them, the sum of their microcosm, that upon which all growth depends, from which all power proceeds. The King in like manner is the foundation of his kingdom, the sun of the world around him, the heart of the republic, the fountain whence all power, all grace doth flow' (W. Harvey, *The Circulation of the Blood*, London, 1906, p. 3).
39. Feyerabend, *Science*, pp. 104–5.
40. C. Lévi-Strauss, *The Savage Mind*, London, 1966, pp. 13–14.
41. Feyerabend, *Science*, p. 102.
42. The issue is discussed by B. Capp in his *Astrology and the Popular Press*, London, 1979, pp. 188–90. Nor have things changed much. The September 1975 issue of the American *Humanist* carried a statement signed by 186 scientists, including 18 Nobel laureates, insisting that 'there is no scientific foundation for [astrology's] tenets.' The arguments offered were purely traditional: the origin of astrology in magic, the immense distance of the stars and planets from the earth, and our ignorance of any astrological mechanism.
43. A. De Grazia, *The Velikovsky Affair*, London, 1978; also of interest is S. L. Talbott (ed.), *Velikovsky Reconsidered*, London, 1978. George Zweig, co-author in 1964 of quark theory, had trouble having his ideas taken seriously: 'When the physics department of a leading university was considering an appointment for me, their senior theorist, one of the most respected spokesmen for all theoretical physics, blocked the appointment at a faculty meeting by passionately arguing that the model was the work of a "charlatan"' (Harald Fritsch, *Quarks*, Harmondsworth, 1984, p. 66).
44. Feyerabend, *Science*, p. 103.
45. Ibid., p. 103n.
46. The claims are taken from an undated *Thorson's Book Catalogue*.
47. B. Inglis, *The Hidden Power*, London, 1986, p. 61.
48. R. S. Worrall, 'Iridology', in K. Frazier (ed.), *Science Confronts the Paranormal*, Buffalo, 1986, pp. 185–97.
49. *Surely You're Joking, Mr Feynman*, London, 1986, pp. 338–9.
50. The story of Kelvin's war against evolution has been fully described by J. Burchfield, *Lord Kelvin and the Age of the Earth*, New York, 1975.
51. M. Gardner, *Science: Good, Bad, and Bogus*, Oxford, 1981, pp. 155–6.
52. Lucretius, *The Nature of the Universe*, Harmondsworth, 1951.
53. Ibid., pp. 36–7.

54. Ibid., pp. 37–8.
55. *Physics*, Book 6, 231.a.25.
56. E. Grant, *Source Book in Medieval Science*, Cambridge, 1974, p. 320.
57. Ibid., p. 316.
58. One of this rare breed, Henry of Harclay, Chancellor of Oxford University in 1312, is quoted by Grant (*Source Book*, pp. 319–20). Aristotelian rather than Lucretian, Harclay's arguments turn upon such issues as whether or not 'God actually perceives or knows the first inchoative point of a line', or whether he can annihilate 'the end point of a line without corrupting the line'.
59. F. Lange, *The History of Materialism*, London, 1950, p. 225.
60. R. H. Kargon, *Atomism in England from Hariot to Newton*, Oxford, 1966, pp. 25–7.
61. M. P. Crosland, *The Science of Matter*, Harmondsworth, 1971, p. 71.
62. H. G. Alexander, *The Leibniz–Clarke Correspondence*, Manchester, 1956, p. 60.
63. Lucretius, *Nature of the Universe*, pp. 71–2.
64. W. Harvey, *On Animal Generation*, Exercise 46.
65. J. Locke, *Essay Concerning Human Understanding*, Book 2, chapter 8, section 23.
66. A good account of the work and influence of Boscovich can be found in L. Pearce Williams, *Michael Faraday*, London, 1965, pp. 73–80.
67. I. Kant, *Critique of Pure Reason*, London, 1934, p. 157.
68. Ibid., p. 157.
69. Pearce Williams, *Faraday*, pp. 376–8.
70. Oxygen was discovered by Priestley and Scheele in 1773, and nitrogen by Rutherford the year before.
71. *Principia*, Book 2, Prop. 23.
72. Crosland, *Science of Matter*, p. 200.
73. Ibid., pp. 201–2.
74. Ibid., pp. 202–4.
75. Ibid., p. 206.
76. D. Knight, *Atoms and Elements*, London, 1967, p. 114.
77. R. D. Hoblyn, *A Treatise on Chemistry*, London, 1860, p. 16.
78. G. F. Rodwell (ed.), *A Dictionary of Science*, London, 1871, p. 58.
79. Crosland, *Science of Matter*, p. 207.
80. E. Mach, *The Science of Mechanics*, La Salle, Ill., 1962, pp. 492–4, 505.
81. Crosland, *Science of Matter*, p. 236.
82. M. J. Nye, *Molecular Reality*, London, 1972, p. 22.
83. The full story is told by M. J. Nye, ibid.
84. Ibid., p. 157.
85. Ibid., p. 161.

86. Ibid., p. 168.
87. *The Scientific Image*, Oxford, 1980.

9 Can Science and Philosophy Conflict?

1. *Cartesian Economics* (London, 1922); *Wealth, Virtual Wealth, and Debt* (London, 1926); *Money versus Men* (London, 1931); *The Role of Money* (London, 1934); *The Archenemy of Economic Freedom* (London, 1943).

2. Some, however, in Nazi Germany were prepared to argue that Einstein's theory of relativity was a function of his Jewishness. Phillipp Lenard, for example, in his *Deutsche Physik* (4 vols, 1936–7), set out to construct a purely Aryan physics free from all Jewish influence.

3. Famous scientists are frequently asked to confirm the obsessions of the less knowledgeable. Francis Crick reported that 'I have even been approached by a well-meaning and enthusiastic modern clergyman in Cambridge, who suggested that there must be some fascinating relationship between DNA and ESP' (*Of Molecules and Men*, Seattle, 1967, pp. 85–6). More confident, if no less obscure, was the claim of Salvador Dali in 1964: 'And now the announcement of Watson and Crick about DNA. This is for me the real proof of the existence of God' (ibid., p. 1).

4. R. Jenkin, *The Reasonableness and Certainty of the Christian Religion*, London, 1700, vol. 2, p. 18.

5. R. I. Watson (ed.), *Basic Writings in the History of Psychology*, New York, 1979, pp. 87–8.

6. K. R. Popper, *The Poverty of Historicism*, London, 1961, p. iv.

7. J. Locke, *An Essay Concerning Human Understanding*, Book 4, chapter 3, section 29.

8. J. L. Austin, *Philosophical Papers*, Oxford, 1970, p. 232.

9. Locke, *Essay*, Book 4, chapter 21.

10. J. S. Mill, *A System of Logic*, Book 6, chapter 3.

11. Aristotle, *On the Parts of Animals*, 652.b.

12. Ibid., 666.a.

13. Plato, *Timaeus*, ed. H. D. P. Lee, Harmondsworth, 1965, pp. 95–9.

14. *The Philosophical Writings of Descartes*, ed. J. Cottingham, R. Stoothoff, and D. Murdoch, Cambridge, 1985, vol. 1, pp. 340–8.

15. A full account of Whytt's work and the ventricular theory can be found in R. K. French's valuable study, *Robert Whytt, the Soul, and Medicine*, London, 1969.

16. Franz Gall published his full system in his *Sur les Fonctions du Cerveau*, Paris, 1822–5, 6 vols (*On the Functions of the Brain*, Boston, 1835, 6 vols).

17. This is something of an oversimplification. Historians distinguish between Gall's views and those of his popularisers. Michael Shortland, for

example, in his 'Courting the Cerebellum', *British Journal for the History of Science*, April 1987, pp. 173–99, distinguished between organology and phrenology. Gall, an organologist, was an anatomist who taught that the brain was indeed a complex organ, the site of all propensities, sentiments and faculties. The phrenological tradition, however, which claimed to read faculties from bumps, derived not from Gall, but from his former collaborator, J. C. Spurzheim (1776–1832).

18. The number of organs varied from book to book; it also tended to grow with the century.

19. S. Shapin, 'The Politics of Observation', in R. Wallis (ed.), *On the Margins of Science*, Keele, 1979, p. 154.

20. F. A. Lange, *The History of Materialism*, London, 1957, Book 2, p. 123.

21. In 1924 the *Times Literary Supplement* spoke of *parapsychology*, referring to the study of phenomena lying outside orthodox psychology. A second term, *paraphysics*, was coined by the physicist Oliver Lodge in 1933 to describe 'physical phenomena for which no adequate scientific explanation exists'. The more general term *parascience* was used for the first time in *Mind* (1953) to describe the study of phenomena 'beyond the scope of scientific enquiry'. *Parascientists* first appeared in *Nature* in 1974. It is presumably only a matter of time before the term *paratruth* appears. Further details can be found in M. Thalbourne and R. Rosenbaum, 'The Origin of the Word "parapsychology"', *Journal of the Society for Psychical Research*, vol. 53, January 1986, pp. 225–9.

22. In 1979 the theoretical physicist John Wheeler unsuccessfully called upon the AAAS to sever its connection with parascience. 'We have enough charlatanism in this country today without needing a scientific organisation to prostitute itself to it', he charged. More details can be found in M. Gardner, *Science: Good, Bad and Bogus*, Oxford, 1981, pp. 185–206.

23. See Alan Gauld, *The Founders of Psychical Research*, London, 1968, pp. 3–31.

24. R. G. Medhurst, *Crookes and the Spirit World*, London, 1972, p. 116.

25. D. Hume, *Enquiries*, ed. L. A. Selby-Bigge, Oxford, 1955, p. 119.

26. Soal's records of tests with Basil Shackleton were shown by B. Marwick to have been systematically doctored: 'The Soal–Goldney Experiments with Basil Shackleton: New Evidence of Data Manipulation', *Proceedings of the Society for Psychical Research*, vol. 56, May 1978, pp. 250–77. In 1974 Levy was observed to interfere with an automatic recorder supposedly being influenced by psychic rats.

27. Burt's first full report on the IQ of separated twins, published in 1955, listed 21 pairs. The further studies dating from 1958 and 1966 raised their number to 'over 30' and 53 pairs respectively. In all cases, Kamin

noted, the correlation between IQ scores of the separated twins remained 0.771 exactly. The probability of such exact correlation occurring naturally is very slight. Further reading of Burt's work confirmed Kamin's suspicion that it was fraudulent.

28. L. J. Kamin, *The Science and Politics of I.Q.*, Harmondsworth, 1974; O Gillie, *Sunday Times*, 24 October 1976; L. S. Hearnshaw, *Cyril Burt*, London, 1979.

29. H. Eysenck and L. J. Kamin, *Intelligence*, London, 1981, p. 99.

30. Medhurst, *Crookes*, pp. 130–41.

31. See R. S. Westfall, 'Newton and the Fudge Factor', *Science*, vol. 179, 1973, pp. 751–8.

32. R. H. Thouless, *Experimental Psychical Research*, Harmondsworth, 1963, p. 82.

33. 'Precognition of a Quantum Process', *Journal of Parapsychology*, vol. 33, 1969, pp. 99–108.

34. Compare on this point Harry Collins's account in *Changing Order* (London, 1985, pp. 51–78) of a colleague's attempt to build a TEA-laser. Despite the fact that the technology of the laser was known, and that Harrison, Collins's colleague, was an experienced laser-builder familiar with this particular design, it still took six months 'from the assembly of the parts to the final ironing out of the faults to make Jumbo work'. After numerous reports of unexpected loud bangs, loud cracks, falling voltages, and capacitors connected the wrong way round, Collins began to suspect that they were running out of things to test. Eventually, Harrison, 'short of anything else to do', removed a connecting wire from the spark gap and Jumbo at last began to work. And all this, Collins emphasised, is a piece of 'straightforward "normal science" where no one doubted that the phenomenon could be replicated'.

35. Or even how often those 'neat little diagrams' were wrong. Watson in *The Double Helix* (London, 1968, pp. 190–1) described how one of his early DNA models had been destroyed by the crystallographer Jerry Donohue: 'The tautomeric forms I had copied out of Davidson's book were, in Jerry's opinion incorrectly assigned . . . Happily he let out that for years organic chemists had been arbitrarily favouring particular tautomeric forms over their alternatives on only the flimsiest of grounds. In fact, organic-chemistry textbooks were littered with pictures of highly improbable tautomeric forms. The guanine picture I was thrusting towards his face was almost certainly bogus.'

36. The papers of this controversy have been reprinted in *Isaac Newton's Papers and Letters on Natural Philosophy*, ed. I. B. Cohen, London, 1978. The story of the French reaction to Newton's work has been vividly told in Henry Guerlac, *Newton on the Continent*, Ithaca, 1981.

37. F. Close, M. Marten, and C. Sutton, *The Particle Explosion*, Oxford, 1987, p. 206.
38. Ibid., p. 204.
39. Ibid., p. 203.
40. R. Feynman, *Surely You're Joking, Mr Feynman*, London, 1986, p. 344.
41. Ibid., pp. 344–5.
42. J. Beloff and D. Bate, 'An attempt to replicate the Schmidt findings', *Journal of the Society for Psychical Research*, vol. 46, 1971, pp. 21–31.
43. John Hasted, *The Metal-Benders*, London, 1981, p. 55.
44. Ibid., p. 2.
45. *Observer*, 8 March 1981.
46. *Metal-Benders*, p. 45.
47. S. G. Soal and H. T. Bowden, *The Mind Readers*, London, 1954.
48. The polywater story in all its fascinating detail has been beautifully told in F. Franks, *Polywater*, Boston, 1983.
49. R. J. Strutt, *Life of Lord Rayleigh*, London, 1924, p. 387.
50. Galileo, *Dialogue on the Great World Systems*, Chicago, 1953, p. 469.
51. Many of these and other scientific lapses can be followed in two recent surveys: W. Broad and N. Wade, *Betrayers of Truth*, London, 1983; A. Kohn, *False Prophets*, Oxford, 1986.
52. C. Bernard, *An Introduction to the Study of Experimental Medicine*, New York, 1957, p. 202.
53. See p. 164 above.
54. A. Koestler, *The Roots of Coincidence*, London, 1974, p. 50.
55. L. Susan Stebbing, *Philosophy and the Physicists*, Harmondsworth, 1944.
56. Cicero, *The Nature of the Gods*, Harmondsworth, 1972, p. 159.
57. Lucretius, *The Nature of the Universe*, Harmondsworth, 1951, p. 177.
58. Cicero, *Nature of the Gods*, pp. 226–35.
59. The correspondence with Bentley can be consulted in *The Correspondence of Isaac Newton*, ed. H. W. Turnbull, vol. 3, Cambridge, 1961, pp. 233–41, 244–5, 253–6.
60. Ibid., p. 233.
61. Ibid., p. 234.
62. Ibid., p. 235.
63. Ibid., p. 244.
64. Ibid., p. 244.
65. Ibid., p. 235.
66. *Pensées*, 77.
67. I. Newton, *Opticks*, New York, 1979, p. 402.
68. See I. Hacking, *The Emergence of Probability*, Cambridge, 1975, pp. 166–71.
69. C. A. Russell (ed.), *Science and Religious Belief*, London, 1973, p. 183.

70. Precisely this point was made by Halley in the 'Ode' he contributed to *Principia*:

> Now we know
> The sharply veering ways of comets, once
> A source of dread, nor longer do we quail
> Beneath appearances of bearded stars.

71. H. G. Alexander (ed.), *The Leibniz–Clarke Correspondence*, Manchester, 1956, pp. 11–14.
72. R. Wollheim (ed.), *Hume on Religion*, London, 1963, pp. 99–204.
73. Ibid., p. 117.
74. Ibid., pp. 116–17.
75. Ibid., p. 123.
76. Ibid., p. 154.
77. The references are to the General Scholium with which Newton concluded *Principia*, and to the Rules of Reasoning in Philosophy which preface Book 3.
78. I. Kant, *Critique of Pure Reason*, London, 1934, p. 365.
79. While Paley could find in the eye 'a cure for atheism', astronomy was dismissed in a few pages. It was not, he argued, 'the best medium through which to prove the agency of an intelligent Creator'. The heavens were too simple, showing nothing 'but bright points'. For design to reveal itself some 'complexity' was needed. Paley did, however, allow that the heavens could reveal, if not the creator's existence, 'the magnificence of his operations' (*Natural Theology*, London, 1846, chapter 22).
80. *Enquiries*, p. 114.
81. C. Babbage, 'Argument in Favour of Design', in I. B. Cohen and H. M. Jones (eds), *Science before Darwin*, London, 1963, pp. 52–76.
82. H. Miller, *Testimony of the Rocks*, Edinburgh, 1871, p. 184.
83. Ibid., p. 185.
84. C. Darwin, *The Origin of Species*, ed. J. W. Burrow, Harmondsworth, 1968, ch. 4.
85. R. Dawkins, *The Blind Watchmaker*, London, 1986.
86. Ibid., p. 21.
87. Freeman Dyson, *Disturbing the Universe*, New York, 1979, p. 251.
88. F. Hoyle, *Galaxies, Nuclei, and Quasars*, London, 1966, p. 150.
89. See J. Barrow and F. Tippler, *The Anthropic Cosmological Principle*, Oxford, 1988, p. 22.
90. Ibid., p. 16.
91. Ibid., p. 3.
92. Ibid., p. 21.

10 Could Science Have Developed Differently?

1. F. Yates, *Giordano Bruno and the Hermetic Tradition*, London, 1964; G. de Santillana, *The Crime of Galileo*, London, 1958. The issue is discussed in two more general works: R. Hooykaas, *Religion and the Rise of Modern Science*, Edinburgh, 1972; R. S. Westfall, *Science and Religion in Seventeenth-Century England*, Ann Arbor, Mich., 1973.

2. J. L. Heilbron, *Elements of Early Modern Physics*, Berkeley, Cal., 1982, pp. 21–2.

3. Erasmus, *In Praise of Folly*, ed. A. H. T. Levi, Harmondsworth, 1971, pp. 151–2.

4. J. B. della Porta, *Natural Magick*, ed. D. de Solla Price, New York, 1957, Book 1, section 8.

5. See Heilbron, *Elements*, p. 19.

6. See p. 109 above.

7. Porta, *Natural Magick*, pp. 191–2.

8. Heilbron, *Elements*, pp. 16–17.

9. Agrippa's work is discussed in Frances Yates, *The Occult Philosophy in Elizabethan England*, London, 1983, pp. 37–47.

10. E. Rosen, *Three Copernican Treatises*, New York, 1959, p. 147.

11. W. Gilbert, *De Magnete*, ed. P. Fleury Mottelay, New York, 1958, pp. 81–2, 103.

12. Ibid., p. 103.

13. Ibid.

14. R. S. Westman, 'Nature, Art and Psyche', in B. Vickers (ed.), *Occult and Scientific Mentalities in the Renaissance*, Cambridge, 1984, p. 205. The Kepler–Fludd dispute is discussed in several of the essays in Vickers (ed.), *Occult and Scientific Mentalities*: in particular, R. S. Westman's 'Nature, Art, and Psyche', pp. 177–230; E. Rosen's 'Kepler's Attitude toward Astrology and Mysticism', pp. 253–72; and J. V. Field's 'Kepler's Rejection of Numerology', pp. 273–96.

15. F. Bacon, *The Advancement of Learning*, ed. A. Johnston, Oxford, 1980, p. 26.

16. D. Lindberg, *Theories of Vision from Al Kindi to Kepler*, Chicago, 1976, p. 98.

17. *The Philosophical Writings of Descartes*, ed. J. Cottingham, R. Stoothoff, and D. Murdoch, Cambridge, 1985, vol. 1, pp. 153–4.

18. *Leibniz: Philosophical Writings*, ed. G. H. R. Parkinson, London, 1973, p. 179.

19. *Selected Philosophical Papers of Robert Boyle*, ed. M. A. Stewart, Manchester, 1979, p. 68.

20. Gilbert, *De Magnete*, p. 102.

21. J. Locke, *An Essay Concerning Human Understanding*, Book 2, chapter 31, section 6.

22. Descartes, *Philosophical Writings*, p. 285.

23. Boyle, *Selected Philosophical Papers*, p. 68.

24. The alchemists favoured the three-element theory of salt, sulphur, and mercury. For other systems see my *The Classics of Science*, New York, 1984, pp. 264-5.

25. R. Boyle, *The Sceptical Chymist*, London, 1911, pp. 167-8.

26. Ibid., pp. 178-9.

27. Ibid., p. 179.

28. Ibid., pp. 182-3.

29. Descartes, *Philosophical Writings*, p. 202.

30. Boyle, *Selected Philosophical Papers*, p. 7.

31. J. Annas and J. Barnes, *The Modes of Scepticism*, Cambridge, 1985, p. 5.

32. R. Popkin, *The History of Scepticism from Erasmus to Spinoza*, Berkeley, Cal., 1979, pp. 37-41.

33. Ibid., pp. 133-4.

34. J. Needham, *Science and Civilisation in China*, vol. 2, Cambridge, 1956, p. 13.

35. Ibid., p. 29.

36. Compare Newton on this issue. See above, p. 66.

37. Needham, *Science and Civilization*, vol. 2, p. 47.

38. Ibid., p. 55.

39. Ibid., p. 71.

40. Ibid., p. 437.

41. N. Sivin, 'The Scientific Revolution in China', in E. Mendelsohn (ed.), *Transformation and Tradition in the Sciences*, Cambridge, 1984, p. 533.

11 Are Scientific Theories Rational? Are They True?

1. Paul Hazard, *The European Mind, 1680-1715*, Harmondsworth, 1964, p. 188.

2. In antiquity Hesiod had spoken of the gold, silver and bronze races of the distant past, while in the medieval period the prophet Joachim of Fiore had divided history into the ages of the Father, Son and Holy Ghost. In the nineteenth century Renan wrote of the three ages of myth, analysis, and complete understanding. More recently the philosopher R. Collingwood in his *Essay on Metaphysics* (Oxford, 1940, pp. 51-5) divided the development of modern science into three phases: Newtonian science with its presupposition that some events have causes, Kantian science based on the assumption that all events have causes, and finally, the science of Einstein with its presupposition that no events have causes. A further division was proposed by the ecologist Aldo Leopold in the 1940s. Initially, he argued, we recognised duties to our

fellow-men. Secondly, we came to recognise that we also had a duty to society. And, finally, we are entering an age in which we will eventually realise our equally pressing duties to the land and the animals and plants dependent upon it. (For more about Leopold see J. Passmore, *Man's Responsibility for Nature*, London, 1980.)

3. Mary Kingsley, writing in 1897, from first-hand experience had somehow managed to persuade herself that African languages were not elaborate enough 'to enable a native to state his exact thought. Some of them are very dependent upon gesture. When I was with the Fans they frequently said "We will go to the fire so that we can see what they say", when any question had to be decided after dark, and the inhabitants of Fernando Po, the Bubis, are quite unable to converse with each other unless they have sufficient light to see the accompanying gestures of the conversation' (*Travels in West Africa*, reprinted London, 1965, p. 504).

4. K. R. Popper, *Conjectures and Refutations*, London, 1969, p. 151.

5. E. E. Evans-Pritchard, *Witchcraft, Oracles and Magic among the Azande*, Oxford, 1937, pp. 63–4.

6. Ibid., p. 65.

7. Ibid., p. 24.

8. Ibid., p. 65.

9. E. Durkheim and M. Mauss, *Primitive Classification*, London, 1903, p. 6.

10. Later investigations have indeed revealed a more prosaic picture. Apparently only male Bororo claim to be parrots; Bororo women merely keep them as pets. The claim of kinship therefore reduces to no more than an expression of male dependence on women in a society governed by rules of matrilineal descent and uxorilocal residence. See J. C. Crocker, 'My brother the Parrot', in J. D. Sapir and J. C. Crocker (eds), *The Social Use of Metaphor*, Philadelphia, pp. 164–92.

11. B. Malinowski, *Argonauts of the Western Pacific*, London, 1922; *Coral Gardens and their Magic*, London, 1935; R. Firth, *We, the Tikopia*, London, 1936; *Primitive Polynesian Economy*, London, 1939.

12. The original paper was first published in two parts in *Africa*, 1967, vol. 37, pp. 50–71, 155–87. It became better known in an abbreviated version published in B. R. Wilson (ed.), *Rationality*, London, 1970, pp. 130–71. See also Horton's later paper, 'Tradition and Modernity Revisited', in M. Hollis and S. Lukes (eds), *Rationality and Relativism*, Oxford, 1982, pp. 201–60.

13. Horton, 1970, p. 132.

14. Ibid., p. 136.

15. S. J. Tambiah, 'Classification of Animals in Thailand', and R. Bulmer, 'Why the Cassowary is not a Bird', in M. Douglas (ed.), *Rules and Meanings*, Harmondsworth, 1973, pp. 127–66, 167–93.

16. Facsimiles of the first (1735) and tenth (1758) editions of Linnaeus's *Systema naturae* have been published in Nieuwkoop (1964) and London (1956). More conveniently the 'Animal Kingdom of Linnaeus' (1735) is presented in English translation in D. R. Oldroyd, *Darwinian Impacts*, Milton Keynes, 1983, pp. 16–17.

17. More details can be found in D. J. Kevles, *In the Name of Eugenics*, Harmondsworth, 1986, pp. 238–41.

18. W. Emerson, *Principles of Mechanics*, London, 1773, cited by L. Laudan in R. Butts and J. Davis (eds), *The Methodological Heritage of Newton*, Oxford, 1970, p. 104n.

19. Evans-Pritchard, *Witchcraft*, pp. 24–5.

20. K. R. Popper, *Conjectures and Refutations*, London, 1969, p. 34.

21. C. Payne-Gaposchkin, *Her Autobiography and other Recollections*, Cambridge, 1984, p. 117.

22. T. Kuhn, *The Structure of Scientific Revolutions*, Chicago, 1970, pp. 76, 77, 78, 35–6.

23. M. Polanyi, *Personal Knowledge*, Chicago, 1958, pp. 286–94.

24. Galileo, *Dialogue Concerning Two New Sciences*, ed. G. de Santillana, New York, 1952, pp. 178ff.

25. A. Koyré, *Metaphysics and Measurement*, London, 1968, p. 94. Yet some of these experiments have been repeated by physicists and have been found to work out exactly as Galileo described them. See T. B. Settle, 'An Experiment in the History of Science', *Science*, vol. 133, 1961, pp. 19–23, and the discussion of the issue by I. B. Cohen, *The Birth of a New Physics*, Harmondsworth, 1987, pp. 85–101, 196–204.

26. J. R. Partington, 'The Origins of the Atomic Theory', *Annals of Science*, vol. 4, 1939, p. 278.

27. R. A. Fisher, 'Has Mendel's Work been Rediscovered?', *Annals of Science*, vol. 1, 1936, pp. 115–37.

28. R. S. Westfall, 'Newton and the Fudge Factor', *Science*, vol. 179, 1973, pp. 751–58; R. R. Newton, *The Crime of Ptolemy*, Baltimore, 1977; G. Holton, 'Subelectrons, Presuppositions, and the Millikan–Ehrenhaft Dispute', *Historical Studies in the Physical Sciences*, vol. 9, 1978, pp. 166–224.

29. The list of as yet undetected particles is in fact quite large. A full list has been compiled by John Kenny, an American physicist, and contains more than 100 named particles. Arranged alphabetically it extends from 'A: alphon, anomalons, arion, axiinstanton, axino, axion' to 'Z: zerino, zeron'. See *New Scientist*, 5 May 1988, pp. 71–2.

30. J. Swift, *Gulliver's Travels*, ed. P. Dixon and J. Chalker, Harmondsworth, 1967, pp. 242–3.

31. H. Poincaré, *Science and Hypothesis*, New York, 1952, p. 160.

32. H. Roscoe, *Lessons in Elementary Chemistry*, London, 1887, pp. 294–5.

33. *International Critical Tables*, ed. E. W. Washburn, 7 vols, New York and London, 1926–33; *Barker Index*, ed. M. W. Porter and R. C. Spiller, Cambridge, 1951; *Index Kewensis*, 4 vols, Oxford, 1893–5, Supplements 1–12, 1901–59.

34. W. Whewell, *History of the Inductive Sciences*, London, 1857, vol. 1, p. 8.

35. I. Hacking, *Representing and Intervening*, Cambridge, 1983, p. 21.

36. Ibid., p. 27.

37. R. Harré, *Varieties of Realism*, Oxford, 1986, p. 95.

38. Loschmidt actually calculated the number of molecules of a gas in 1 cm^3 at normal temperatures and pressures. This number equals $2.687 \times 10^{19}/cm^3$. As Avogadro's number is defined as the number of molecules in 1 gram-molecular volume, 22.4 litres for a gas at normal temperature and pressure, it must equal Loschmidt's number × 22,400.

39. M. J. Nye, *Molecular Reality*, London, 1972, pp. 16ff.

40. W. Salmon, *Scientific Explanation and the Causal Structure of the World*, Princeton, NJ, 1984, pp. 213–27.

41. L. Laudan, *Science and Values*, Berkeley, Cal., 1984, pp. 113–16.

42. Bas C. van Fraasen, *The Scientific Image*, Oxford, 1980, p. 12.

43. Merton's first major work in the field was *Science, Technology and Society in Seventeenth-Century England*, first published as an article in 1938 in *Osiris* (vol. 4, pt. 2) and republished in 1970 (New York). Much of his later work has been collected in his *The Sociology of Science*, Chicago, 1973.

44. B. Barnes and D. Bloor, 'Relativism, Rationalism, and the Sociology of Knowledge', in M. Hollis and S. Lukes (eds), *Rationality and Relativism*, Oxford, 1982, p. 23.

45. Ibid., p. 34.

46. M. Hollis, 'Reason and Ritual', in B. R. Wilson (ed.), *Rationality*, Oxford, 1970, pp. 221–39. The laws mentioned by Hollis are the law of excluded middle (p is either true or false), the law of contradiction (p is not both true and false), and *modus ponens* (if p is true, and if p implies q, then q is also true).

47. Barnes and Bloor, 'Relativism', p. 45.

48. See A. R. Anderson and N. Belnap, 'Tautological Entailments', *Philosophical Studies*, 13, 1962, pp. 9–24.

49. Barnes and Bloor, 'Relativism', p. 46.

50. See J. Haberer, *Politics and the Community of Science*, New York, 1969, pp. 103–62; and S. J. Gould, *The Mismeasure of Man*, Harmondsworth, 1984.

51. A. Pannekoek, *A History of Astronomy*, London, 1961, p. 230.

INDEX